Praise for Wallace J. Nichols's

BLUE

"A fascinating study of the emotional, behavioral, psychological, and physical connections that keep humans so enchanted with water." —Nicola Joyce, *Washington Post*

"Nichols's strength is inspiring people to remember why they care for the oceans." —Erik Vance, *Conservation*

"Using a combination of anecdotes and hard data, Nichols makes a persuasive case for water's healing power."

—Amanda FitzSimons, *Elle*

"Nichols insists on a more experiential brand of activism, encouraging individuals to encounter the oceans directly in order to build personal and emotional connections—to get their feet wet." —*GQ*

"*Blue Mind* has the power to reveal how your health and happiness are tied to oceans and waterways, no matter where on Earth you live."

—Dr. Sylvia Earle, founder of Mission Blue and Explorer in Residence at *National Geographic*

"Nichols shares the many ways in which water positively impacts not just our lives, but also our minds. With many citations for the scientifically curious, and numerous anecdotes that entertain, *Blue Mind* will interest a broad audience, from fans of water sports to ecologists."

—*Publishers Weekly*

"The implications of this premise are profound; they may revolutionize the way we teach conservation and ecology."

—Jeff Greenwald, *OnEarth*

"Every conservation movement requires sound science and effective communication. Neuroconservation skillfully utilizes the insight of neuroscience to accomplish both of these goals. Already, Nichols's work is sending ripples throughout the world."

—Will Fraker, *Bay Nature*

"A fascinating, fact-based report for aquaphiles and those at one with the tides."

—*Kirkus Reviews*

"It is impossible not to engage with Nichols as an author; his irrepressible sense of wonder and passion for his subject are simply inspiring."

—Kelsy Peterson, *Library Journal*

"If your time on our planet needs less stress and more happiness, find your way to Nichols's *Blue Mind* for the surprisingly simple and joyous ways water can change your life."

—Timothy Dykman, director, Ocean Revolution

"Nichols wants us to learn to see the ocean differently, and not just because it will make us happier, but because by reimagining it we become able to reimagine ourselves, and by extension our relationship with the world we inhabit."

—James Bradley, *The Australian*

"Nichols celebrates the spiritual connection that water provides.... He digs into meaning and metaphor.... Empathy is central to his theme.... I liked this book for its mad spirit."

—Philip Hoare, *The Guardian*

"Beyond the physical gains you'll make, there is also a burgeoning body of research that proves proximity to water can reduce stress and improve mood."

—Devin Tomb, *Self*

BLUE MIND

The Surprising Science That Shows How Being Near, In, On, or Under Water Can Make You Happier, Healthier, More Connected, and Better at What You Do

WALLACE J. NICHOLS

FOREWORD BY CÉLINE COUSTEAU

Little, Brown Spark
New York • Boston • London

Little, Brown Spark
Hachette Book Group
1290 Avenue of the Americas, New York, NY 10104
littlebrownspark.com

Originally published in hardcover by Little, Brown and Company, July 2014
First Little, Brown Spark paperback edition, July 2015

Little, Brown Spark is an imprint of Little, Brown and Company, a division of Hachette Book Group, Inc. The Little, Brown Spark name and logo are trademarks of Hachette Book Group, Inc.

ISBN 978-0-316-25208-9 (hc) / 978-0-316-37148-3 (int'l ed) / 978-0-316-25211-9 (pb)
LCCN 2014937382

11

LSC-C

Printed in the United States of America

For my mothers and fathers,
(S, S, K, W, J, V & J)

brothers and sisters,
(J, J, J, J, J, P, R, E & P)

daughters,
(G & J)

and my beloved,
(D)

I wish you water.
(J)

Contents

Foreword

Céline Cousteau

Of my grandfather's many famous quotes, these are the two I hear, read, and see emblazoned on walls and websites most often:

> "The sea, once it casts its spell, holds one in its net of wonder forever."
>
> "People protect what they love."

What did he mean when he said them? I can only infer—he's not here to ask—but I believe they should be taken in simplicity. The first is about that magical enchantment so many people feel toward the water. The second is about survival—protecting what we love is a basic instinct to humans (just ask any parent). Taken together, these two ideas explain the lives of many aquatic scientists: you are pulled in and fall in love with the water world, so you dedicate yourself to protecting it. And yet, in an effort to remain unbiased, neutral, and impartial, these same scientists wouldn't think to investigate what's behind the passion that drives their work.

The same is true for millions of nonscientists who choose to go to the water's edge for a vacation. Most don't consider why this is where they best relax, replenish, and rejuvenate.

They don't question that they're getting something cognitively enriching from being by the water that goes way beyond sitting in a beach chair with a best-selling novel. They just know they love and need unplugged waterside time.

I'm not so different. I was born into an ocean family; my grandfather Jacques helped bring the wonders of the undersea world to millions via television, and this is part of my DNA, part of how I function. But there's also a part of me that doesn't want to know why I love the water, a part that prefers to just see its effects as magical, something unknown—indeed, unknowable—but deeply felt.

I don't think everything in life needs to be explained. But when I consider what's at stake, and the fact that we must encourage people to change their behavior and shift government policies if we're going to restore the health of the world's water systems, I've come around to agreeing with my dear friend Dr. Wallace "J." Nichols that it's time to explain the magic.

While appealing to people's emotions can be very effective in many instances, at some point you have to be able to appeal to their intellects. You also need to adapt the message to different audiences. If you're talking to a lawmaker, a fisher, a surfer, a real estate agent, or a mom, you have to speak their language. This sometimes means backing up the awe and wonder with numbers and facts, biology and neurochemistry.

As this book demonstrates, neuroscientists and psychologists are now producing research data that accomplish exactly that. We are beginning to learn that our brains are hardwired to react positively to water and that being near it can calm and connect us, increase innovation and insight, and even heal what's broken. Healthy water is crucial to our physiological and psychological well-being, as well as our ecology and economy. We have a "blue mind"—and it's perfectly tailored to make us happy in all sorts of ways that go far beyond relaxing in the surf, listening to the murmur of a stream, or floating quietly in a pool.

Because our love of water is so pervasive, so consistent, it can seem that asking why is a question that needs no answer. But once you begin to go deeper (no pun intended!), things are not as simple as we might initially think. We love the rhythmic sound of waves breaking on the beach, but why does *that* sound relax us more than nearly any other? How might our preference for the flat surface of a lake be rooted to prehistoric hunting patterns? Why does understanding the science of somatic tension help explain the pleasure we feel when submerged? And on and on, as you'll see in this book. The result is not just understanding but inspiration. A comparison might be to gravity: we know what it *is*, but if our curiosity had stopped simply at the point of realizing that what we toss into the air must come down, we could never have landed men on the moon. So, too, water's effects on us.

Fortunately, as I've become more involved with J.'s Blue Mind project, I've come to realize that understanding the science behind our feelings for water doesn't do anything to diminish those feelings. As J. likes to say, "Looking at the science of wonder and love doesn't make it any less wonderful." Of course, statements like that can make scientists uncomfortable. Many people are afraid of the "touchy-feely" stuff that comes along with exploring human emotions. In 2013, when I attended the third annual Blue Mind Summit on Block Island, I found myself in an isolated place surrounded by people from all walks of life—neuroscientists, psychologists, educators, divers, artists, musicians—and water. We provoked each other and explored different methods of explaining universal sensations. I go to a lot of forums and conferences, and usually we check our personal feelings and experiences with nature at the door, reserving them strictly for the closing keynote remarks, or limiting them to one-on-one exchanges during breaks. The neuroscientists who gave presentations had to explain their work in the context of this multidisciplinary event and tell us how it applies to emotion. A few of them said while they were comfortable writing reports full of charts, data, brain imagery, and chemical formulas, they didn't quite know how to *speak* universally about their science. But when they did, it resonated with everyone, because all of a sudden those of us who don't usually understand that side of things thought, "Oh my God, really? That's what happens? My neurons are doing *that*? It really is your brain on water!"

I strongly believe that the results of all of this research, debate, discussion, and celebration should be shared widely. Former boundaries are being pushed—even shoved—into new territory. More people must be included in the exchange until the wisdom that emerges from this conversation becomes common knowledge among members of the human tribe. Blue Mind is, deep down, about human curiosity, knowing ourselves more and better. If J. weren't curious, we wouldn't be where we are now.

The sea, once it casts its spell, does indeed hold us in its net of wonder forever. People do protect what they love. Why are these the two simple ideas we repeat so often? Because they're true. Now it's time for us to update my grandfather's prescient words by explaining them, and, by doing so, changing our understanding and our consciousness of our place on the planet, who we are, and what goes on inside and between us. It's about reconnecting our sense of self and soul with our waterways and oceans. It's about finding creativity, clarity, and confidence in our deep Blue Minds.

My grandfather would go there, and so shall we.

Preface

One of the many possible ways to describe a life would be as a series of encounters with various bodies of water. Time spent in, on, under, or near water interspersed with the periods spent thinking about where, when, and how to reach it next.

My first body of water, of course, was experienced as a zygote in my mother's womb. And the last—at least as I now imagine it—will be in the form of ashes, cast over the Pacific.

In between, I've been fascinated by and privileged to know many ponds, tanks, rivers, bottles, pools, lakes, streams, buckets, waterfalls, quarries, tubs, mists, oceans, downpours, and puddles.

As children we delight in water. As we grow older, water also becomes the matrix for sport, relaxation, and romance.

My parents took me to the Caribbean as a small child. The photos from that trip seem so familiar that I can still feel the day: sitting on the beach next to the ocean, smiling in the Bahamian sun. I believe my happy memories of the sea were carried forward by those cherished, faded photographs.

Soon after that trip, prior to my third birthday, I had a vivid dream in anticipation of a celebration. At the party in my dream we all sat at a round table under the peach tree in my backyard in Westwood, New Jersey. Everyone received a gift. We were served tea, and at the bottom of the teacups were iron figurines. Somehow, we each became very small and the cups became enormous as we dove down to the bottom to find and retrieve our gift. My friend Steve got a race car. Rusty's was a dog. Mine was a black bear standing on all four legs. I loved that dream—so much so that I tried to dream it again every night before going to sleep. And every time I saw a bear, or a cast-iron car or dog, or a cup of tea, I thought of my dream. That went on for months and then years, dreaming and daydreaming, and wanting to dream about diving into a teacup to retrieve an iron bear. I still have that dream.

At five years of age I became more curious about being adopted. Questions just seemed to lead to more questions and eventually a driven inquiry into the basics of human genetics. That same year I was afflicted with a severe case of spinal meningitis and hospitalized. It was then that I also became intimately familiar with—and curious about—my own nervous system. My adoptive mother was a nurse, and texts and manuals from nursing school days became the scriptures of my childhood. Science, exploration, medicine, and the existence of occupations related to helping people heal grew as a seed in my mind.

In high school, my favorite weekend activity was to push off the shore at night in a canoe with just a box of Pop-Tarts, a fishing pole, and Pyotr Ilich Tchaikovsky. Fish or no fish, the solitude of night drifting was an epic escape.

A few years later, in college at DePauw University, I began to wonder more formally as a young student of science about why I liked water so much. Snorkeling in Bowman Pond on campus and scuba diving in Indiana's quarries were somewhat unusual activities. Exploring the many creeks, rivers, and lakes of the Midwest, I also began to explore the human brain, somewhat by accident.

My sophomore year I was invited by the university chaplain to provide guitar lessons as a volunteer at a nursing home in town. I obliged and ended up spending Wednesday afternoons for eight months playing music with Barbara Daugherty, a woman who had lost her memory—including her ability to play guitar—in an automobile accident fifteen years prior, when she herself was a sophomore in college. The music lessons seemed to trigger long-lost memories, which, once flowing, often continued. The nurses were impressed. I was too, and curious. I'd return to campus and try to learn more about the brain–music connection from professors and journals, without much luck. These days a Google search would turn up endless publications on the topic, but in 1986 that wasn't the case. This early brush with music therapy was logged deep in my memory.

At Duke University I explored wild rivers and the Outer Banks and studied economics, public policy, and decision science. But our science and policy texts were incapable of including the feeling of running a rapid, sitting at the bottom of a quarry, the physiology of retrieved memories and nostalgia, or the creative elixir of floating beneath the stars to *Swan Lake*.

After receiving a doctorate from the landlocked University of Arizona, I proposed to my wife, Dana, underwater off the coast of Cabo Pulmo, Mexico. I wordlessly slipped a sea-turtle-shaped ring onto her finger.

As parents of Grayce and Julia, our favorite moments together involve water.

After two decades working as a marine biologist studying sea turtles, the brain-on-water theme remained on my mind. In fact, curiosity about neuroscience often informed our approach to rebuilding sea turtle populations, one human at a time.

In 2009, the Pew Marine Fellows program generously nominated me for one of their annual awards, as they had a few years prior. The first time I had proposed a community-based sea turtle research project. This time I proposed looking into the science behind our emotional connection to water. I figured that if the pull of water could guide my life so far—as well as that of many, if not most, of my colleagues—those emotions might also be worth knowing more about.

As a non-neuroscientist I composed a rather good proposal about Blue Mind and submitted it to the foundation. The first round (sea turtles) I had been denied the fellowship because I was "too young." This time the response was "too creative." Despite these setbacks, both projects have moved forward, and I have greatly enjoyed the many collaborations and contributions that have come from them.

Now I am neither too young nor too creative, but I am patient, persistent, and truly enamored.

This book is the result of that mix: a life driven by a love of water, some patience and persistence, and a lot of collaboration and conversation with fellow water lovers and scientists, a truly excellent group of people.

Near the end of *The Ocean of Life,* marine biologist Callum Roberts's thorough and insightful treatment of the history of ocean use and overuse, he describes some of the fundamental ingredients needed for fixing what's broken on our blue planet: "It is essential for ocean life and our own that we transform ourselves from being a species that uses up its resources to one that cherishes and nurtures them."

The same can be said for our planet's lakes, rivers, and wetlands, as well as its forests and prairies.

But if this is the emotional foundation of our future, insights into what it means to *cherish* and *nurture* could be useful indeed. How do these Blue-Minded emotions work? What are they made of, and how do we make more? Those are some of the fundamental questions of neuroconservation.

Roberts continues, "People have a deep emotional connection to the sea. The oceans inspire, thrill, and soothe us. Some think we owe our clever brains and the success they brought to our ancestors' close link to the sea. But our relationship with the sea stretches back through time much further than this: all the way to the origins of life itself. We are creatures of the ocean."

Clearly, creating more protection and restoration will require that we better appreciate and understand the science behind, and what goes into, the mysterious elixir called *inspiration*, the chemistry of *thrill,* and the main ingredient found in *soothe.*

Combined with pinches of empathy, nostalgia, responsibility, gratitude, and a big scoop of love for our waters, we have a fighting chance to get this right.

> *You have to do it because you can't stand not to. That's the best reason to do anything.*
>
> — Linda Ronstadt

Truth be known, I tried my hardest to give this project away to those with better training, better brains, better résumés for the job. There were no takers. So, I built upon what I have learned about people and water from my teachers: Herman Melville, Joshua Slocum, Chuy Lucero, Don

Thomson, Loren Eiseley, Jacques Cousteau, Pak Lahanie, Wade Hazel, Pablo Neruda, Juan de la Cruz Villalejos, Sylvia Earle, Mike Orbach, Cecil Schwalbe, and Mary Oliver.

Mostly, we've connected the dots that we could find and worked to make the best sense of the patterns that emerged. The goal has been less about providing absolute answers and more about asking new questions—questions that, hopefully, in your capable hands, lead to creative new ways of exploring living well together on our water planet.

Back in 2005, the late author David Foster Wallace opened his commencement speech at Kenyon College with a story about three fish: "There are these two young fish swimming along and they happen to meet an older fish swimming the other way, who nods at them and says 'Morning, boys. How's the water?' And the two young fish swim on for a bit, and then eventually one of them looks over at the other and goes, 'What the hell is water?' "

That's the fundamental question that started my own journey. What is water? Why are we humans so enthralled by it? And why is this question so obvious and important, yet so hard to adequately answer?

Later in his speech, Wallace told the graduating class that education should be based on simple awareness: "Awareness of what is so real and essential, so hidden in plain sight all around us, all the time, that we have to keep reminding ourselves over and over: 'This is water.' "

This book is an attempt to begin a conversation about water based on new questions and current research. I hope to bring to our simple awareness the reality and essence—and beauty—of this small blue marble we live, move, and love upon.

Even though it's hidden in plain sight.

BLUE MIND

1

Why Do We Love Water So Much?

Water is life's matter and matrix, mother and medium. There is no life without water.

— Albert Szent-Györgyi, M.D.,
discoverer of vitamin C

I'm standing on a pier at the Outer Banks of North Carolina, fifty feet above the Atlantic. To the left and right, forward, back, and below, all I can see is ocean. I'm wearing a light blue hat that looks like a bejeweled swim cap, and a heavy black cable snakes down my back like a ponytail. Even though I look like an extra from an Esther Williams movie who wandered into Woody Allen's *Sleeper* by mistake, in truth I'm a human lab rat, here to measure my brain's response to the ocean.

The cap is the nerve center of a mobile electroencephalogram (EEG) unit, invented by Dr. Stephen Sands, biomedical science expert and chief science officer of Sands Research. Steve's a big, burly, balding guy of the sort that could be mistaken for the local high school science teacher who's also the football coach, or perhaps the captain of one of the deep-sea fishing boats that call the Outer Banks home. An El Paso (a city on the Rio Grande) resident by way of Long Beach, California, and Houston, Texas, Steve spent years in academia as a professor, using brain imaging to research Alzheimer's disease. In 1998 he established Neuroscan, which became the largest supplier of EEG equipment and software for use in neurological research. In 2008 Steve founded Sands Research, a company that does neuromarketing, a new field using behavioral and neurophysiological data to track the brain's response to advertising. "People's responses to any kind of stimulus, including advertising, include conscious activity—things we can verbalize—and subconscious activity," he once wrote. "But the subconscious responses can't be tracked through traditional market research methods." When groups of neurons are activated in the brain by any kind of stimulus—a picture, a sound, a smell, touch, taste, pain, pleasure, or emotion—a small electrical charge is generated, which indicates that neurological functions such as memory, attention, language processing, and emotion are taking place in the cortex. By scrutinizing where those electrical charges occur in the brain, Steve's sixty-eight-channel, full-spectrum EEG machine can measure everything from overall engagement to cognition, attention, the level of visual or auditory stimulation, whether the subject's motor skills are involved, and how well the recognition and memory circuits are being stimulated. "When you combine EEG scans with eye-movement tracking, you get unique, entirely nonverbal data on how someone is processing the media or the real-world environment, moment by moment," Steve says.

Given current perplexity about the value of promotional

efforts, Steve's data are increasingly sought after. Sands Research does advertising impact studies for some of the largest corporations in the world; it's perhaps best known for an "Annual Super Bowl Ad Neuro Ranking," which evaluates viewers' neurological responses to those $3.8-million-per-thirty-second spots. (Among those that Steve's team measured were the well known ads that featured people sitting on a beach, backs to the camera as they gazed at white sand and blue water, Corona beers on the table between them, and only the lapping of the sea as a soundtrack. That campaign made the brewer famous, forever associated with tropical ocean leisure.)

In the months prior to my trip to the Outer Banks, I'd been contacted by Sands Research's director of business development, Brett Fitzgerald. Brett's an "outside" kind of guy with a history of working with bears in Montana. He'd heard about my work combining water science with neuroscience and contacted me to see if we could do some sort of project together. Before I knew it, he was on a plane to California, and we met along the coast north of my home to talk "brain on ocean." Not long after, I was on a plane heading to North Carolina.

Today Brett has fitted me with a version of the Sands Research EEG scanning apparatus that can detect human brain activity with the same level of precision as an fMRI (functional magnetic resonance imaging). The data from the electrodes in this ornamented swim cap are sampled 256 times per second and, when amplified for analysis, will allow neuroscientists to see in real time which areas of the brain are being stimulated. Typically such data are used to track shoppers' responses in stores like Walmart as they stop to look at new products on a shelf. In this case, however, the sixty-eight electrodes plugged into the cap on my head are for measuring my every neurological up and down as I plunge into the ocean. It's the first time equipment like this has been considered for use at (or in) the water, and I'm a little anxious about both the current incompatibility (no pun intended) between the technology and the ocean, but also about what we might learn. So is Brett—the cap and accompanying scanning device aren't cheap. In the future such a kit will be made waterproof and used underwater, or while someone is surfing. But for today, we're just hoping that neither the equipment nor I will be the worse for wear after our testing and scheming at the salt-sprayed pier.

It's only recently that technology has enabled us to delve into the depths of the human brain *and* into the depths of the ocean. With those advancements our ability to study and understand the human mind has expanded to include a stream of new ideas about perception, emotions, empathy, creativity, health and healing, and our relationship with water. Several years ago I came up with a name for this human–water connection: Blue Mind, a mildly meditative state characterized by calm, peacefulness, unity, and a sense of general happiness and satisfaction with life in the moment. It is inspired by water and elements associated with water, from the color blue to the words we use to describe the sensations associated with immersion. It takes advantage of neurological connections formed over millennia, many such brain patterns and preferences being discovered only now, thanks to innovative scientists and cutting-edge technology.

In recent years, the notion of "mindfulness" has edged closer and closer to the mainstream. What was once thought of as a fringe quest for Eastern vacancy has now been recognized as having widespread benefits. Today the search for the sort of focus and awareness that characterizes Blue Mind extends from the classroom to the boardroom to the battlefield, from the doctor's office to the concert hall to the world's shorelines. The stress produced in our overwhelmed lives makes that search more urgent.

Water's amazing influence does not mean that it displaces other concerted efforts to reach a mindful state; rather, it adds to, enhances, and expands. Yet this book is not a field guide to meditation, nor a detailed examination of other means toward a more mindful existence. To use a water-based metaphor, it offers you a compass, a craft, some sails, and a wind chart. In an age when we're anchored by stress, technology, exile from the natural world, professional suffocation, personal anxiety, and hospital bills, and at a loss for true privacy, casting off is wonderful. Indeed, John Jerome wrote in his book *Blue Rooms* that "the thing about the ritual morning plunge, the entry into water that provides the small existential moment, is its total privacy. Swimming is between me and the water, nothing else. The moment the water encloses me, I am, gratefully, alone." Open your Blue Mind and the ports of call will become visible.

To properly navigate these depths, over the past several years I've brought together an eclectic group of scientists, psychologists, researchers, educators, athletes, explorers, businesspeople, and artists to consider a fundamental question: what happens when our most complex organ—the brain—meets the planet's largest feature—water?

As a marine biologist as familiar with the water as I am with land, I believe that oceans, lakes, rivers, pools, even fountains can irresistibly affect our minds. Reflexively we know this:

there's a good reason why Corona chose a beach and not, say, a stockyard. And there are logical explanations for our tendency to go to the water's edge for some of the most significant moments of our lives. But *why?*

I look out from the pier at the vast Atlantic and imagine all the ways that the sight, sound, and smell of the water are influencing my brain. I take a moment to notice the feelings that are arising. For some, I know, the ocean creates fear and stress; but for me it produces awe and a profound, immersive, and invigorating peace. I take a deep breath and imagine the leap, cables trailing behind me as I plunge into the waves surging around the pier. The EEG readings would reflect both my fear and exhilaration as I hit the water feet first. I imagine Dr. Sands peering at a monitor as data come streaming in.

Water fills the light, the sound, the air—and my mind.

Our (Evolving) Relationship to Water

Thousands have lived without love, not one without water.

—W. H. Auden

There's something about water that draws and fascinates us. No wonder: it's the most omnipresent substance on Earth and, along with air, the primary ingredient for supporting life as we know it. For starters, ocean plankton provides more than half of our planet's oxygen. There are approximately 332.5 million cubic miles of water on Earth—96 percent of it saline. (A cubic mile of water contains more than 1.1 trillion gallons.) Water covers more than 70 percent of Earth's surface; 95 percent of those waters have yet to be explored.

From one million miles away our planet resembles a small blue marble; from one hundred million miles it's a tiny, pale blue dot. "How inappropriate to call this planet Earth when it is quite clearly Ocean," author Arthur C. Clarke once astutely commented.

That simple blue marble metaphor is a powerful reminder that ours is an aqueous planet. "Water is the *sine qua non* of life and seems to be all over the universe and so it's reasonable for NASA to use a 'follow the water' strategy as a first cut or shorthand in our quest to locate other life in the universe," Lynn Rothschild, an astrobiologist at the NASA Ames Research Center in Mountain View, California, told me. "While it may not be the only solvent for life, it certainly makes a great one since it is abundant, it's liquid over a broad temperature range, it floats when solid, allowing for ice-covered lakes and moons, and it's what we use here on Earth."

Whether searching the universe or roaming here at home humans have always sought to be by or near water. It's estimated that 80 percent of the world's population lives within sixty miles of the coastline of an ocean, lake, or river. Over half a billion people owe their livelihoods directly to water, and two-thirds of the global economy is derived from activities that involve water in some form. Approximately a billion people worldwide rely primarily on water-based sources for protein. (It's very possible that increased consumption of omega-3 oils from eating fish and shellfish played a crucial role in the evolution of the human brain. And, as we'll discuss later in the book, the seafood market is now global in a manner that could never have been imagined even a few decades ago.) We use water for drinking, cleansing, working, recreating, and traveling. According to the U.S. Geological Survey, each person in the United States uses eighty to one hundred gallons of water every day for what we consider our "basic needs." In 2010 the General Assembly of the United Nations declared, "Safe and clean drinking water is a human right essential to the full enjoyment of life."

Our innate relationship to water goes far deeper than economics, food, or proximity, however. Our ancient ancestors came out of the water and evolved from swimming to crawling to walking. Human fetuses still have "gill-slit" structures in their early stages of development, and we spend our first nine months of life immersed in the "watery" environment of our mother's womb. When we're born, our bodies are approximately 78 percent water. As we age, that number drops to below 60 percent—but the brain continues to be made of 80 percent water. The human body as a whole is almost the same density as water, which allows us to float. In its mineral composition, the water in our cells is comparable to that found in the sea. Science writer Loren Eiseley once described human beings as "a way that water has of going about, beyond the reach of rivers."[1]

We are inspired by water—hearing it, smelling it in the air, playing in it, walking next to it, painting it, surfing, swimming or fishing in it, writing about it, photographing it, and creating lasting memories along its edge. Indeed, throughout history, you see our deep connection to water described in art, literature, and poetry. "In the water I am beautiful," admitted Kurt Vonnegut.[2] Water can give us energy, whether it's hydraulic, hydration, the tonic effect of cold water splashed on the face, or the mental refreshment that comes from the gentle, rhythmic sensation of hearing waves lapping a shore. Immersion in warm water has been used for millennia to restore the body as well as the mind. Water drives many of our decisions—from the seafood we eat, to our most romantic moments, and from where we live, to the sports we enjoy, and the ways we vacation and relax. "Water is something that humanity has cherished since the beginning of history, and it means something different to everyone," writes archeologist Brian Fagan. We know instinctively that being by water makes us healthier, happier, reduces stress, and brings us peace.

In 1984 Edward O. Wilson, a Harvard University biologist, naturalist, and entomologist, coined the term "biophilia" to describe his hypothesis that humans have "ingrained" in our genes an instinctive bond with nature and the living organisms we share our planet with. He theorized that because we have spent most of our evolutionary history—three million years and 100,000 generations or more—in nature (before we started forming communities or building cities), we have an innate love of natural settings. Like a child depends upon its mother, humans have always depended upon nature for our survival. And just as we intuitively love our mothers, we are linked to nature physically, cognitively, and emotionally.

> *You didn't come into this world. You came out of it, like a wave from the ocean. You are not a stranger here.*
>
> —Alan Watts

This preference for our mother nature has a profound aesthetic impact. The late Denis Dutton, a philosopher who focused on the intersection of art and evolution, believed that what we consider "beautiful" is a result of our ingrained linkage to the kind of natural landscape that ensured our survival as a species. During a 2010 TED talk, "A Darwinian Theory of Beauty," Dutton described findings based on both evolutionary psychology and a 1997 survey of contemporary preference in art. When people were asked to describe a "beautiful" landscape, he observed, the elements were universally the same: open spaces, covered with low grass, interspersed with trees. And if you add water to the scene—either directly in view, or as a distant bluish cast that the eye takes as an indication of water—the desirability of that landscape skyrockets. Dutton theorized that this "universal landscape" contains all the elements needed for human survival: grasses and trees for food (and to attract edible animal life); the ability to see approaching danger (human or animal) before it arrives; trees to climb if you need to escape predators; and the presence of an accessible source of water nearby. In 2010 researchers at Plymouth University in the United Kingdom asked forty adults to rate over one hundred pictures of different natural and urban environments. Respondents gave higher ratings for positive mood, preference, and perceived restorativeness to *any* picture containing water, whether it was in a natural landscape or an urban setting, as opposed to those photos without water.

Marcus Eriksen, a science educator who once sailed a raft made entirely of plastic bottles from the U.S. Pacific coast to Hawaii, expanded upon Dutton's hypothesis to include seacoasts, lakeshores, or riverbanks. In the same way the savannah allowed us to see danger a long way off, he theorized, coastal dwellers could see predators or enemies as they came across the water. Better, land-based predators rarely came from the water, and most marine-based predators couldn't emerge from the water or survive on land. Even better than that: the number of food and material resources provided in or near the water often trumped what could be found on land. The supply of plant-based and animal food sources may vanish in the winter, Eriksen observed, but our ancestors could fish or harvest shellfish year-round. And because the nature of water is to move and flow, instead of having to travel miles to forage, our ancestors could walk along a shore or riverbank and see what water had brought to them or what came to the water's edge.

While humans were developing an evolutionary preference for a certain type of water-containing landscape, the human *brain* was also being shaped by environmental demands. Indeed, according to molecular biologist John Medina, the human brain evolved to "solve problems related to surviving in an unstable outdoor environment, and to do so in nearly constant motion." Imagine that you are one of our distant *Homo sapiens* ancestors, living in that ideal savannah landscape more than 200,000 years ago. Even if you and your family have inhabited this particular spot for a while, you still must be alert for any significant threats or potential sources of food. Every day brings new conditions—weather, animals, fruits, and other edible plants. Use up some sources of food and you have to look for more, which means constant exploration of your environment to learn more about where you are and what other sources of food and water are available for you and your family. Perhaps you encounter new plants or animals, some of which are edible—some not. You learn from your mistakes what to gather and what to avoid. And while you and your children learn, your brains are being shaped and changed by multiple forces: your individual experiences, your social and cultural interactions, and your physical environment. Should you survive and reproduce, some of that rewiring will be passed on to your descendants in the form of a more complex brain. Additional information for survival will be socially encoded in vivid stories and songs.

A nervous system is the part of an animal that coordinates activity by transmitting signals about what's happening both inside and outside the body. It's made up of special types of cells called neurons, and ranges in size and complexity from just a few hundred nerve cells in the simplest worms, to some 20,000 neurons in the California sea hare, *Aplysia californica* (a very cool mollusk whose large, sometimes gigantic, neurons have made it the darling of neurobiologists for the past fifty years), to as many as 100 billion in humans. We'll be looking in detail at the human brain and DNA in later chapters, but there's an important point to be made before we leave our ancestors on the distant savannah: just as the human brain changed and evolved over the millennia, our *individual* brain changes and evolves from the day we are born until we die. Critical studies starting in the 1970s and 1980s demonstrated that our brains are in a state of constant evolution—neurons growing, connecting, and then dying off. Both the brain's physical structure and its functional organization are *plastic*, changing throughout our lives depending on need, attention, sensory input, reinforcement, emotion, and many other factors. The brain's *neuroplasticity* (its ability to continually create new neural networks, reshape existing ones, and eliminate networks that are no longer used due to changes in behavior, environment, and neural processes) is what allows us to learn, form memories throughout our lifetimes, recover function after a stroke or loss of sight or hearing, overcome destructive habits and become better versions of ourselves. Neuroplasticity accounts for the fact that, compared to most of us, a disproportionate amount of physical space in a violinist's brain is devoted to controlling the fingers of his or her fingering hand, and that studying for exams can actually increase the amount of cortical space devoted to a particular subject (more complex functions generally require more brain matter). As we'll see later, it also accounts for certain negative behaviors, like obsessive-compulsive disorder.

You will hear the term neuroplasticity a lot in this book, because it exemplifies one of the fundamental premises of Blue Mind: the fact that our brains—these magnificent, three-pound masses of tissue that are almost 80 percent water—are shaped, for good or ill, by a multitude of factors that include our perceptions, our emotions, our biology, our culture—and our environment.

You'll also hear a lot about happiness. While the "pursuit of happiness" has been a focus of humankind since almost before we could put a name to the feeling, from ancient times onward philosophers have argued about the causes and uses of happiness, and composers, writers, and poets have filled our heads with stories of happiness lost and found. In the twenty-first century, however, the pursuit of happiness has become one of the most important means of judging our quality of life. "Happiness is an aspiration of every human being," write John F. Helliwell, Richard Layard, and Jeffrey D. Sachs in the United Nations' *World Happiness Report 2013*, which ranks 156 countries by the level of happiness of their citizens.[3] It's a vital goal: "People who are emotionally happier, who have more satisfying lives, and who live in happier communities, are more likely both now and later to be healthy, productive, and socially connected. These benefits in turn flow more broadly to their families, workplaces, and communities, to the advantage of all."[4]

"The purpose of our lives is to be happy," says the Dalai Lama—and with all the many benefits of happiness, who would disagree? As a result, today we are bombarded with books on happiness, studies (and stories) about happiness, and happiness research of every kind. We'll walk through some of the studies later, and discuss why water provides the most profound shortcut to happiness, but suffice it to say, greater individual happiness has been shown to make our relationships better; help us be more creative, productive, and effective at work (thereby bringing us higher incomes); give us greater self-control and ability to cope; make us more charitable, cooperative, and empathetic;[5] boost our immune, endocrine, and cardiovascular systems; lower cortisol and heart rate, decrease inflammation, slow disease progression, and increase longevity.[6] Research shows that the amount of happiness we experience spreads outward, affecting not just the people we know but also the friends of their friends as well (or three degrees of the famous six degrees of separation).[7] Happy people demonstrate better cognition and attention, make better decisions, take better care of themselves, and are better friends, colleagues, neighbors, spouses, parents, and citizens.[8] Blue Mind isn't just about smiling when you're near the water; it's about smiling *everywhere.*

Water and Our Emotions

> *Some people love the ocean. Some people fear it. I love it, hate it, fear it, respect it, resent it, cherish it, loathe it, and frequently curse it. It brings out the best in me and sometimes the worst.*
>
> — Roz Savage

Beyond our evolutionary linkage to water, humans have deep *emotional* ties to being in its presence. Water delights us and inspires us (Pablo Neruda: "I need the sea because it teaches me"). It consoles us and intimidates us (Vincent van Gogh: "The fishermen know that the sea is dangerous and the storm terrible, but they have never found these dangers sufficient reason for remaining ashore"). It creates feelings of awe, peace, and joy (The Beach Boys: "Catch a wave, and you're sitting on top of the world"). But in almost all cases, when humans think of water—or hear water, or see water, or get in water, even taste and smell water—they feel *something.* These "instinctual and emotional responses... occur separately from rational and cognitive responses," wrote Steven C. Bourassa, a professor of urban planning, in a seminal 1990 article in *Environment and Behavior.* These emotional responses to our environment arise from the oldest parts of our brain, and in fact can occur before any cognitive response arises. Therefore, to understand our relationship to the environment, we must understand both our cognitive *and* our emotional interactions with it.

This makes sense to me, as I've always been drawn to the stories *and* science of why we love the water. However, as a doctoral student studying evolutionary biology, wildlife ecology, and environmental economics, when I tried to weave emotion into my dissertation on the relationship between sea turtle ecology and coastal communities, I learned that academia had little room for feelings of any kind. "Keep that fuzzy stuff out of your science, young man," my advisors counseled. Emotion wasn't rational. It wasn't quantifiable. It wasn't science.

Talk about a "sea change": today cognitive neuroscientists

have begun to understand how our emotions drive virtually every decision we make, from our morning cereal choice, to who we sit next to at a dinner party, to how sight, smell, and sound affect our mood. Today we are at the forefront of a wave of neuroscience that seeks to discover the biological bases of everything, from our political choices to our color preferences. They're using tools like EEGs, MRIs, and those fMRIs to observe the brain on music, the brain and art, the chemistry of prejudice, love, and meditation, and more. Daily these cutting-edge scientists are discovering why human beings interact with the world in the ways we do. And a few of them are now starting to examine the brain processes that underlie our connection to water. This research is not just to satisfy some intellectual curiosity. The study of our love for water has significant, real-world applications—for health, travel, real estate, creativity, childhood development, urban planning, the treatment of addiction and trauma, conservation, business, politics, religion, architecture, and more. Most of all, it can lead to a deeper understanding of who we are and how our minds and emotions are shaped by our interaction with the most prevalent substance on our planet.

The journey in search of people and scientists who were eager to explore these questions has taken me from the sea turtles' habitats on the coasts of Baja California, to the halls of the medical schools at Stanford, Harvard, and the University of Exeter in the United Kingdom, to surfing and fishing and kayaking camps run for PTSD-afflicted veterans in Texas and California, to lakes and rivers and even swimming pools around the world. And everywhere I went, even on the airplanes connecting these locations, people would share their stories about water. Their eyes sparkled when they described the first time they visited a lake, or ran through a sprinkler in the front yard, caught a turtle or a frog in the creek, held a fishing rod, or walked along a shore with a parent or boyfriend or girlfriend. I came to believe that such stories were critical to science, because they help us make sense of the facts and put them in a context we can understand. It's time to drop the old notions of separation between emotion and science—for ourselves and our future. Just as rivers join on their way to the ocean, to understand Blue Mind we need to draw together separate streams: analysis and affection; elation and experimentation; head and heart.

The Tohono O'odham (which means "desert people") are Native Americans who reside primarily in the Sonoran Desert of southeastern Arizona and northwest Mexico. When I was a graduate student at the University of Arizona, I used to take young teens from the Tohono O'odham Nation across the border to the Sea of Cortez (the Gulf of California). Many of them had never seen the ocean before, and most were completely unprepared for the experience, both emotionally and in terms of having the right gear. On one field trip several of the kids didn't bring swim trunks or shorts—they simply didn't own any. So we all sat down on the beach next to the tide pools of Puerto Peñasco, I pulled out a knife, and we all cut the legs off our pants, right then and there.

Once in the shallow water we put on masks and snorkels (we'd brought enough for everyone), had a quick lesson on how to breathe through a snorkel, and then set out to have a look around. After a while I asked one young man how it was going. "I can't see anything," he said. Turns out he'd been keeping his eyes closed underwater. I told him that he could safely open his eyes even though his head was beneath the surface. He put his face under and started to look around. Suddenly he popped up, pulled off his mask, and started shouting about all the fish. He was laughing and crying at the same time as he shouted, "My planet is beautiful!" Then he slid his mask back over his eyes, put his head back into the water, and didn't speak again for an hour.

My memory of that day, everything about it, is crystal clear. I don't know for sure, but I'll bet it is for him, too. Our love of water had made an indelible stamp on us. His first time in the ocean felt like mine, all over again.

The Beginnings of Blue Mind

In 2011, in San Francisco—a city surrounded by water on three sides—I gathered a group of neuroscientists, cognitive psychologists, marine biologists, artists, conservationists, doctors, economists, athletes, urban planners, real estate agents, and chefs to explore the ways our brains, bodies, and psyches are enhanced by water. I had realized that there was a constellation of innovative thinkers who had been trying to put the pieces together regarding the powerful effects of water, but they had mostly been isolated from one another. Since then, the Blue Mind gathering has become an annual conference that taps into a growing quantity of new mind/body/environment research and continues to produce new and startling insights on how humanity interacts with our watery planet. Both the brain and the ocean are deep, complex, and subtle realms—scarcely explored and poorly understood. However, we are on the cusp of an age when both the brain and the ocean are giving up more and more of their secrets to dedicated scientists and explorers. As more researchers from varied disciplines apply their expertise to the relation between water and humanity, the insights from their collaborations are illuminating the biological, neurological, and sociological benefits of humanity's Blue Mind.

Every year more experts of all kinds are connecting the dots between brain science and our watery world. This isn't touchy-feely "let's save the dolphins" conservation: we're talking prefrontal cortex, amygdala, evolutionary biology, neuroimaging, and neuron functioning that shows exactly why humans seem to value being near, in, on, or under the water. And this new science has real-world implications for education, public policy, health care, coastal planning, travel, real estate, and business—not to mention our happiness and general well-being. But it's science with a personal face; science practiced by real people, with opinions, biases, breakthroughs, and insights.

At subsequent Blue Mind conferences on the shores of the Atlantic and Pacific, scientists, practitioners, and students have continued to share their research and life's work, huddling together to discuss, create, and think deeply. We've produced documents describing "what we think we know" (facts), "what we want to explore" (hypotheses), and "what we want to share" (teachings). At Blue Mind 2013, held on Block Island, we discussed topics like dopaminergic pathways, microplastics and persistent organic pollutants, auditory cortex physiology, and ocean acidification, but for those of us drawn to the waves, no discussion of water is without joy and celebration. At dawn we sang together, overlooking the sparkling blue

Atlantic Ocean, and in the evening we drank wine, those waters now black and sparkling, and listened to former Rhode Island poet laureate Lisa Starr.

> *"Listen, dear one," it whispers.*
> *"You only think you have*
> *forgotten the impossible.*
>
> *"Go now, to that marsh beyond*
> *Fresh Pond and consider how the red*
> *burgeons into crimson;*
> *go see how it's been preparing forever*
> *for today."*

This is poetry, this is science; this is science, this is poetry. So, too, are oceans and seas, rivers and ponds, swimming pools and hot springs—all of us could use a little more poetry in our lives.

We could use a *lot* more, too—and, in some cases, a lot less. Too many of us live overwhelmed—suffocated by work, personal conflicts, the intrusion of technology and media. Trying to do everything, we end up stressed about almost anything. We check our voice mail at midnight, our e-mail at dawn, and spend the time in between bouncing from website to website, viral video to viral video. Perpetually exhausted, we make bad decisions at work, at home, on the playing field, and behind the wheel. We get flabby because we decide we don't have the time to take care of ourselves, a decision ratified by the fact that those "extra" hours are filled with e-mailing, doing reports, attending meetings, updating systems to stay current, repairing what's broken. We're constantly trying to quit one habit just to start another. We say the wrong things to people we love, and love the wrong things because expediency and proximity make it easier to embrace what's passing right in front of us. We make excuses about making excuses, but we still can't seem to stop the avalanche. All of this has a significant economic cost as "stress and its related comorbid diseases are responsible for a large proportion of disability worldwide."[9]

It doesn't have to be that way. The surfers, scientists, veterans, fishers, poets, artists, and children whose stories fill this book know that being in, on, under, or near water makes your life better. They're waiting for you to get your Blue Mind on too.

Time to dive in.

2

Water and the Brain: Neuroscience and Blue Mind

The brain and the ocean are profoundly complex and subtle realms for investigation. We struggle to identify the approaches that yield deeper understanding. We're drawn to their mysteries—and the inherent rhythms that define both—and we strive to find a language to describe them.

— Dr. David Poeppel, professor of psychology and neural science, New York University

Surfer João De Macedo waits on his board, three hundred feet or so from the beach. He's relaxed yet alert, scanning the water for indications of the next wave. As he spots a smooth swell pattern that indicates a potentially rideable wave, anticipation releases another wave—of neurochemicals cascading throughout his brain and body. The water rises up in front of him, and he's instinctively standing on his board, using the brain that has been shaped by the experience of the thousands of waves he's ridden before to look for the perfect entry point. Dopamine explodes over his neurons as he drops into the pocket. He's enclosed in a watery tunnel of cool, blue-green light, surrounded by the smell and the sound of the wave, feeling the air rushing past him as he makes minute physical adjustments to keep the ride going for as long as possible. Feel-good neurotransmitters—adrenaline, dopamine, endorphins—rise in waves inside his brain and body. He shoots out of the pocket just as the wave crest crashes behind him. He smiles, maybe even laughs. Then he turns back over the top of the wave, drops down onto his board, and starts to paddle out as he looks for the next wave—and the next dopamine rush.

The smile on João's face wordlessly declares the thrill and pleasure he's just experienced. Via such facial expressions, and through poetry, literature, and testimonies of all sorts, we humans have been self-reporting the effects of water on our minds and bodies. But it's only in the past two decades or so[1] that scientists have been able to examine what's going on in the human brain when we encounter different aspects of our world and ourselves. Today neuroscience is exploding with studies that track, in infinitesimal detail, what our brains are doing when we are eating, drinking, sleeping, working, texting, kissing, exercising, creating, problem solving, playing.... If humans do it, it would seem that someone is studying it, to find out exactly which circuits are firing and which neurotransmitters are cascading during that particular activity.

Some refer to this new era of the brain as the "Golden Age of Neuroscience." But despite all the studies that make for great news copy (and the colorful brain scans that make us more likely to accept the conclusions they accompany),[2] any scientist worth his or her salt will admit that we are just dipping our collective toes into understanding the processes of the human brain. Back in 2011, eminent neuroscientist V. S. Ramachandran stated that our current understanding of the brain "approximates to what we knew about chemistry in the 19th century—in short, not much."[3] That remains the case.

However, even that "not much" is far more than we have understood for two thousand years about the way the mind works. Undoubtedly many of the conclusions and studies done today will be proven inaccurate ten, twenty, fifty years from now. That's science: you carefully observe the world around you, state a hypothesis, develop experiments or studies to prove or disprove your premise, use the technology you have available to you at the moment, and then develop conclusions based on the results. What's truly exciting about this latest round of investigation, however, is that it endeavors to discover the physiological, chemical, and structural processes that underlie humanity's very subjective experience of the world. What happens in your brain when you see the face of a loved one or gaze out over the rail of a boat? Which circuits are active when you're at your most creative, or when an addict is seeking a fix? Does the environment physically shape our brains just as our brain filters, configures, and interprets our moment-to-moment perceptions of that environment? And is there a way we can use this understanding of how our brains work to help us be happier, more creative, more loving, and less stressed?

As I mentioned earlier, while neuroscience is studying an incredibly broad range of human behaviors and emotions, in every setting possible, for some reason our interaction with water seems to be left out. (Try this: go to your nearest bookstore or library and scan the indexes of any of the popular books on neuroscience, psychology, or self-help for topics related to water.) So, in this chapter we are going to start the conversation at the beginning by examining several major aspects of your brain on water. But first, let's talk about the fundamentals of just how we study the brain.

How We Study the Brain

> *Everything we do, every thought we've ever had, is produced by the human brain. But exactly how it operates remains one of the biggest unsolved mysteries, and it seems the more we probe its secrets, the more surprises we find.*
>
> — Neil deGrasse Tyson, astrophysicist

Humans have always been infinitely curious about what goes on inside our skulls. In the late nineteenth and early twentieth centuries, scientists and theorists such as Sigmund Freud and William James described how we think and feel based upon the subjective experiences reported by their patients, combined with their clinical observation of human behavior. (Even today, self-reporting of subjective experience is vital in studying the ways our brains work.) Medical doctors, too, were a rich source of information; in fact, prior to the twentieth century, our information on the brain mostly arose from studying what happened when it malfunctioned due to disease or injury. But it's one thing to theorize about how brains function normally by seeing what happens when they don't function properly, and another to actually observe the brain in action when it is thinking, sleeping, feeling, creating, or interacting with the outside world. How do we go about understanding the *normal* human brain?

Cue the development of noninvasive techniques and devices

that allowed scientists to track the workings of the normal human brain. The earliest of these devices was the electroencephalography (EEG) machine. Based on the understanding that living tissue has electrical properties, the first use of EEG on human beings occurred in 1924. Over the course of the twentieth century EEG recordings were used as diagnostic tools as well as in research.

An EEG works because neurons in the brain generate small electrical charges when they become active, and when groups of neurons fire together, they create an electrical "wave" that can be detected and recorded. Data are gathered by placing the EEG's electrodes (often embedded in a cap, net, or band) on the head and monitoring the peaks and valleys of electricity generated in the brain. (The signal is amplified for the purposes of analysis.) EEGs can track brain activity by location, showing which side of the brain is involved in a "cognitive event"; by type of brain wave (alpha, beta, theta, and delta, each corresponding to a distinct frequency range and corresponding level of brain activity, making EEGs vital in sleep studies); and by abnormal activity (as is seen in epilepsy, a disease that creates patterns of spikes in electrical activity in the brain). Sophisticated EEG devices can noninvasively sample sixty-eight channels of data every four milliseconds or less, and record electrical events as brief as one millisecond.[4]

Cognitive neuroscientists have found that the EEG can be an extremely useful tool in tracking brain functions like attention, emotional responses, how we retain information, and so on.[5] And in an exciting development for those of us who do research outside of a laboratory setting, EEG readers are becoming smaller and more portable — some even resembling the kind of headset a computer gamer might use. However,

EEG readings indicate electrical activity only at a shallow depth, and many critical functions occur much deeper in the brain. To explore those, other tools were needed. In the last fifty years MRI (good old-fashioned magnetic resonance imaging), positron emission tomography (PET), and single-photon emission computed tomography (SPECT) have been used to produce images of activity deep in the brain by tracking changes in blood flow or metabolic activity.[6] But whereas MRI machines rely solely on magnetic fields and radio waves, PET and SPECT scans use injected radioactive isotopes, which limits their usefulness. A new answer arrived in the 1990s with functional magnetic resonance imaging, or fMRI.

Different areas of the brain become active at different times, depending on the tasks required. More activity requires more oxygen, causing increased blood flow to those areas of the brain. Like their older siblings, fMRI machines use powerful magnetic fields to align the protons in the hydrogen atoms in the blood, and then knock them out of alignment with radio waves. An MRI looks for differences in signals from the hydrogen atoms to distinguish between different types of matter. As the protons realign, they send out different signals for oxygenated and deoxygenated blood—and those signals are what the fMRI reads. As a test subject undertakes an activity—squeezing a hand, for example, or looking at a particular picture—fMRI scans measure the ratio of oxygenated to deoxygenated blood, or blood-oxygen-level-dependent (BOLD) contrast, in different areas of the brain at that moment. The machine's computer then uses a sophisticated algorithm to interpret the data received from the fMRI and represent the contrast ratios in the form of infinitesimal, three-dimensional units called voxels. Different colors are used to indicate the intensity of the energy in that particular area, red being the most intense, purple or black indicating low or no activity. The brighter the color on the scan, the greater the activity in that particular region of the brain, giving rise to the term "lights up" when referring to activated brain regions.

Over the past twenty years fMRI has become the preferred method for measuring brain function, utilized by cognitive scientists, neurologists, neurobiologists, psychologists, neuro-economists, and others.[7] But even though fMRI is one of the best tools we have currently for measuring the brain function (and one of the only tools we have for examining structures deep inside the cranium), it's important also to acknowledge its limitations. First, fMRIs are based on an indirect indicator of brain function. The functioning of the brain is ultimately chemical and electrical in nature: neurons produce electrical messages that are conveyed from one to the other either through direct contact, synapse to synapse, or via chemical neurotransmitters. This activity requires oxygen, which is provided by blood flow to the active areas of the brain; fMRI scans measure that blood flow, not actual neuronal activity. Thus, while fMRIs can tell us which areas of the brain are active, they cannot reveal the specific activating factors. Second, while fMRIs have excellent spatial resolution, showing the location of brain activity within two to three millimeters, because blood flow is far slower than neuronal activity (with a delay of at least two to five seconds between activation of neurons and increased blood flow to the area), the *temporal* resolution of an fMRI scan is much longer than the amount of time needed for the majority of perception or other cognitive processes. (In contrast, EEGs have poorer spatial resolution but, as mentioned earlier, can track electrical charges as fast as one millisecond.) There are also issues with differences in fMRI machines and the complex algorithms required to process the data, as well as variation in the size of voxels (which while tiny are far larger than the neurons they represent).

The most important limitation currently for fMRI-based studies of the brain is that they can track responses of subjects only in a lab rather than in the environments where a particular cognitive activity would logically take place. Imagine you were one of those undergraduates who volunteered to be a subject in a study that used fMRI scans.[8] You would be asked to report to a lab and told that you should leave anything metal at home because of the powerful magnets in the machine. A friend who has previously had an fMRI tells you to dress warmly, as the temperature in the facility is kept cold for the equipment. Once you enter the lab and check in, you're ushered into the room with the scanner: a large, doughnut-shaped machine with a hole in the middle just big enough for a human body. (If you're at all claustrophobic, you're uncomfortable merely looking at the small size of the opening.) The technician instructs you to lie down on a plastic bedlike slab with your head pointing toward the opening in the doughnut. She tells you that the bed will move forward so your head and shoulders are inside the scanner. There will be a mirror above you that will allow you to see a computer screen with instructions for tasks to perform while the scanner takes pictures of the activity in your brain. She gives you earplugs and explains that the scanner is very noisy, and points out a buzzer you can push if at any point you become too uncomfortable to continue. She puts a pillow under your head and one on each side of it to keep it stationary. "Stay as still as you can while the scan is in progress," she requests, and then she slides your head and shoulders into the scanner.

So far, the process is equivalent to an MRI. But instead of just lying in the tube, you look up at the mirror and, as the thump-thump-thump sound of the fMRI commences, you can see the computer screen above you come to life. You follow the instructions and push buttons on the pad in your hand in response to the pictures flashing in front of your eyes. (In the future your fingers may get a rest thanks to eye-tracking technology.) The tasks keep you busy enough that you notice the small size of the space you're inhabiting only a couple of times (which is good, because you came very close to pressing the buzzer that would let the technician know you were ready to quit). At the end of the test the computer screen goes blank and the noise stops—finally. The tech rolls you out of the machine, thanks you for your time, and asks you to schedule another session for the following week. You're cold, your bladder is full, and you have the beginnings of a headache from the noise of the scanner—but it's all for science, right? So you agree to come back.

An fMRI can reveal a great deal about the functioning of the brain, but it can tell us far less about the ways the human brain interacts with the real world. It can scan us while we look at pictures of people who are happy or sad or fearful or angry, but it cannot track our actual interactions with people on the street. It can reveal brain activity when we are calculating mathematical problems or choosing between this food or that beverage, but it cannot yet scan us while we enjoy that crisp, red apple picked straight from the tree or that glass of chardonnay in front of a roaring fireplace—let alone snorkeling above a coral reef. As cognitive neuroscientist and expert on auditory cognition, speech perception, and language comprehension Dr. David Poeppel notes, "Most of the mapping of the brain since the 1990s has been done using fMRI, and the goal is to make some kind of cartography. That's laudable, but having a map is not an explanation. Having a map is just the beginning of the problem." So while EEG and fMRI may be our best current tools to study the brain, and their integrated use a true "best of both worlds" solution, scientists like me, whose interests lie with areas that can't be tested by a static machine in a sterile lab, are looking ahead to new techniques: diffusion tensor imaging, or DTI, which uses the diffusion of water through the brain to track neuronal axon bundles (the "cables" that connect regions) throughout the brain's white matter, thereby showing how information travels through the brain;[9] optogenetics, in which light-sensitive genes can be inserted into neurons to activate or silence them in an instant, allowing researchers to determine a particular neuron's function;[10] even wearable fNIRS (functional near-infrared spectroscopy) headgear[11] that will allow scientists to take readings of people in actual situations and environments.

Still, when it comes to understanding the human brain on water today, we can look at several specific streams of valuable information. First, we start with self-reported experience: How do people feel when they are around water? What effects do they notice? Second, we can take studies that have been done about the cognitive effects of nature as a whole and ask whether the response to water is any different. Third, we can look at the wide range of discovery in cognitive neuroscience, neurobiology, environmental psychology, and neurochemistry, and ask, "Does this apply to the brain on water?"

But we'll start with a very basic question that's actually not so basic at all: what exactly does the brain do?

What Does the Brain Do?

It's a metaphor for how the brain is organized.

— Ray Kurzweil, author, inventor, futurist, when asked why he loves the ocean

The three-pound mass of fat and protein that sits at the top of the spinal cord is not only mostly water, but itself rests in a kind of "water": clear, colorless cerebrospinal fluid—composed of living cells (including immune cells that eliminate infectious pathogens from the nervous system), glucose, protein, lactate, minerals, and water—that cushions the brain from injury, maintains pressure in the cranium at a constant level, and provides enough buoyancy to reduce the brain's effective weight from 1,400 grams to between 25 and 50 grams (thus preventing it from putting too much pressure on its lower levels and cutting off its blood supply).[12] By weight, the brain is the biggest consumer of the body's energy, using approximately 20 to 25 percent of its oxygen and 60 percent of its glucose for communication between neurons and cell-health maintenance.[13] It contains approximately 1.1 trillion cells, of which anywhere from 85 to 100 billion[14] are neurons (what we call "gray matter"), and most of the rest are axons and glia ("white matter"). The glia perform metabolic support functions such as wrapping the conducting axons in myelin (a whitish protein and lipid sheath for nerves) and recycling neurotransmitters. But the anatomy doesn't really tell us much about what the brain actually *does*.

When it comes to interpreting those vast complexities of the human brain, Howard Fields of the University of

California, San Francisco, is an excellent guide. Gray-haired and grandfatherly, with an infectious smile, he's not only a world-class researcher but also a brilliant explainer. While major new discoveries about neurons are made almost every year, Fields told me, the best description of their purpose is as encoders of perception and action. Neuronal cell bodies, he said, are electrochemical, digital on-off switches connected by their axonal "wires" in an intricate network inside your head. There are billions of neurons in the human brain and each one can connect with tens of thousands of others, making *trillions* of connections. By interconnecting, groups of neurons create neural networks that produce every conscious and unconscious impulse, response, or thought you may have: from your ability to sense an itch (and scratch your nose in response), to the "ordinary and exceptional mental activities of attending, perceiving, remembering, feeling, and reasoning."[15] They also trigger the cascades of neurochemicals that mediate our emotions and behaviors in response to stress. By tracking different neural networks in different individuals, neuroscientists can create topographic maps of the brain, although this mapping can be exceedingly difficult, as there are often multiple networks performing different parts of the same function—for instance, the network responsible for the hand bringing food to your mouth is different from the network that directs the more subtle and delicate finger movements used when you hold a pen and write.[16] The complexity of these networks is hinted at in sheer quantity—our neuron count exceeds the closest primate's by an order of magnitude—but in the end it's the quality of cognition and dexterity that most amazes.[17]

How can the brain make sense of the vast "storm" of

perceptions and other stimuli that are flooding our senses every moment? And ultimately, what was the brain slated to do, based upon evolution? According to John Medina, the primary purpose of the human brain is to "(1) solve problems (2) related to surviving (3) in an unstable outdoor environment, and (4) to do so in nearly constant motion."[18] This purpose required the brain to constantly grow and adapt to the challenges humanity encountered in an ever-changing and dangerous world. Through the millennia the human brain evolved in fits and starts, with cognitive functions added and pruned based upon their survival benefits. (Every brain still evolves in the same way, gaining certain cognitive functions from birth, and having others eliminated for efficiency through the years.)[19] What resulted over countless centuries was a brain structure with a flexible architecture, neuroplasticity, and, ultimately, neural networks that were able to acquire the basic building blocks of sight, hearing, smell, and sound, which in turn evolved into such higher-level functions as writing, speech, art, and music.[20]

Actually, my using "ultimately" above isn't quite accurate, since in theory while there are physical limits, there are no final endpoints to evolutionary adaptation. If we're around a million years from now, who knows what we'll look like and be able to do? But undoubtedly for most of us the factors that create evolutionary advantages or disadvantages differ from what they were a few hundred thousand years ago. It doesn't take a Ph.D. to recognize that in addition to the environments in which we find ourselves, we are molded by the mental processes going on inside us and our interactions with other people. Indeed, at every moment we are being inundated with input from several sources: the thoughts streaming through our heads; the bodily sensations that rarely attract our attention (unless they demand it due to pain or pleasure); the specific perceptual and neurochemical torrent produced when another human being enters our focus; and the seemingly overwhelming, never-ending streams of stimuli that arrive from every aspect of the world around us.

How in the world can the brain make sense of all this input? How does it separate the signals that are necessary for survival from all of the other perceptual "noise"? It does so by becoming expert in *pattern recognition* and *prediction*. "The human brain is an amazing pattern-detecting machine," writes psychologist David Pizarro. "We possess a variety of mechanisms that allow us to uncover hidden relationships between objects, events, and people. Without these, the sea of data hitting our senses would surely appear random and chaotic."[21] Consciously—but primarily unconsciously—we scan all of the incoming perceptual data and match it with what we have experienced in the past.[22] The brain focuses on what it deems important (either because it matches previous patterns, or, more often, because it does *not* match what is expected and could possibly be dangerous). It interprets what that information means based on prior experience, and then predicts the consequences of what the information means. The processes of the unconscious mind, unlike the conscious "free decisions," occur automatically and are not available to introspection, discussion, approval, or real-time modification.

These predictions often happen below the level of, and much faster than, conscious thought.[23] As David Eagleman, a neuroscientist at Baylor University's College of Medicine, writes in *Incognito: The Secret Lives of the Brain,* "billions of specialized mechanisms operate below the radar—some collecting sensory data, some sending out motor programs, and the majority doing the main tasks of the neural workforce: combining information, making predictions about what is coming next, making decisions about what to do now."[24] It takes 450 milliseconds for a baseball to reach home plate once it's released from the pitcher's hand. As it zips through those sixty feet of airspace, a tremendous amount of data must get crunched by the batter. Foremost, the swing/no-swing decision must be made, not to mention decisions about bat speed and angle, real-time adjustments related to microclimates, and contemplations of the movement of the rest of the players on the field. Considering that it takes about 200 milliseconds to conclude whether a player should swing his bat, there's precious little time to consciously bring one's full knowledge and experience to bear on a proper analysis of the situation. To even consider those 200 milliseconds "time" at all is generous. Yet since 1901 Major League Baseball's batters have made favorable decisions better than one out of four times they've taken an at bat. Those faring better often take their teams to the playoffs and on to the World Series championship game.

Given the challenge of hitting a 95-mile-per-hour fastball, there's effectively no possibility that sort of average is the result of chance, so clearly some sort of cognition is going on. But how? Once the brain has taken in and analyzed data, it then must decide what to do—whether some kind of action is required. Yet how much data can one receive in a matter of milliseconds? To a great extent, this entire process is based upon trial and error. Our brains make a kind of high-speed cost-benefit computation before we take any action, and then revise that computation based on the outcome. Familiar situations reinforce current neural pathways; but every new bit of data, every new circumstance, every mistake and new action, forces the brain to remodel itself, if only slightly. If you've swung a bat tens of thousands of times, you've had moments when you are conscious about adjustments you're making, such as when the team batting coach offers instruction, and are unconscious about the tens of thousands of times your deeper neural networks have absorbed those millisecond experiences. At a certain point, reliance on your conscious mind can become a disadvantage: think too much, and suddenly you've lost your rhythm and the ball misses the hole, the shot clanks off the rim, you play the wrong note—and you strike out.

Throughout our lives, the brain is literally changing itself, creating networks of neurons that accomplish needed functions efficiently.[25] As Michael Merzenich, the "father" of neuroplasticity, states, "The cerebral cortex... is actually selectively refining its processing capacities to fit each task at hand." This is an incredibly dynamic process, where gray matter volume increases in regions where more neuronal activity occurs.[26] It is also a competitive process in which existing networks that are no longer being used become weaker (as anyone who has learned another language and then attempted to take it up again years later knows).

According to Merzenich, neuroplasticity is essentially bimodal: it strengthens neural networks for the things we pay attention to, and weakens the areas we use the least. More complex skills, such as playing a musical instrument or driving a stick-shift car, bring together different neural networks throughout the brain. And even more intriguingly, abilities lost through injury, as in a stroke, can be regained as the brain rewires itself and redirects neural function to new pathways.[27]

While there are specific types of neural networks that most brains have in common, each person's brain maps are unique.[28] (Tracing these neural networks is the next frontier of the Connectome Project, an effort funded by NIH and spearheaded by a consortium including Washington University, University of Minnesota, Oxford University, and others to map the connections in the human brain in a manner similar to that of the successful Human Genome Project of the 1990s, which mapped human DNA. In this case scientists use brain imaging and a new technique called optogenetics, involving the insertion of special molecules into the neurons of the brain that allow their function to be switched on and off by light.) Importantly, when it comes to Blue Mind, there are neural networks that are shaped by your interaction with your environment—from the time your brain starts to form in the environment of the womb, until you close your eyes on your deathbed. And because of neuroplasticity, we have the opportunity to reshape our brains throughout our lifetimes by changing the input and the environment we choose.

Understanding the power of Blue Mind requires taking a journey from a swimming pool to a cancer ward, from the Australian coast to the inner city, from the corner of a five-million-dollar laboratory to the corner office of a five-billion-dollar company. But, like all journeys, it makes sense to start at home—the home in our head and the home with our bed.

3

The Water Premium

On the beach, you can live in bliss.

— Dennis Wilson of the Beach Boys

In 2003, when our daughter Grayce was only eighteen months old, my wife, Dana, and I took her on a 112-day trek, 1,200 miles on foot, along the coastal trail from Oregon to Mexico.[1] While her memories are certainly faint, she still talks about that trip, and I can see how it has shaped her connection to nature and deepened our father-daughter bond. As we walked the beach, the three of us would pass moms and dads of all creeds and colors chasing laughing toddlers down the sand; young men and women, wet-suited and long-haired, eagerly carrying boards into the waves; hale and hearty elders leaning forward, striding against the wind; thirtysomethings walking different breeds of dog, watching as their pets occasionally dashed into the surf and then came running back; kids of all ages beachcombing or running in and out of the ocean, squealing at the cold waves, splashing their friends. I strolled quietly behind fishers in sweatshirts, standing or sitting in low plastic chairs, rods in hands or stabbed into the sand, lines in the water, buckets of bait by their sides; I detoured around beach towels occupied by sunbathers, eyes closed, luxuriating in doing absolutely nothing at all. And on almost every face I saw happiness—everything from elation to blissful laughter to quiet smiles of inward contentment. I could feel those states reflected in my own as I walked the 1,200-mile boundary of land and ocean.

In 2003, close to the end of my trek along the western coast, I came upon a house for sale. It was right on the beach in Del Mar, California, an upscale community with a small number of homes on the sand. I could see streets and hills crowded with larger houses on bigger lots farther back from the water but with spectacular views of the vast Pacific; this one, however, was tiny: a ramshackle, one-story, 800-square-foot bungalow, with one, maybe two tiny bedrooms, and almost no space between it and the (much) larger homes on either side. By the side of the house facing the beach there was a post, at the top of which was a Plexiglas holder containing information about the property. I opened up the holder and pulled out a flyer. The asking price for this 800-square-foot teardown was $6.3 million.

People want to be by the water and increasingly have been willing to pay more to do it. But why? Why would anyone be willing to pay that much for a sliver of land and a ramshackle house in an upscale, yet not that unusual, small town in California, when they could go five blocks inland, pay half as much for a house with twice the square footage, and walk or ride their bike to the ocean every day? (In 2013 the estimated value of that same property was $9.3 million—still one of the best deals in the neighborhood.) Why would someone pay $120,000 for a strip of land 1 foot wide and 1,885 feet long between two properties in East Hampton, New York?[2] Why do people in California, North Carolina, Massachusetts, Florida, and communities on rivers like the Mississippi put up with erosion, minimal square footage, the possibility of being wiped out by storm or flood, or the slow, steady undermining of the sand by the ocean tides or river currents? And what are they really paying for?

The story of that one little bungalow on the beach in Del Mar is a demonstration of the value of proximity: people want to see and hear the water from where they eat and sleep—and they are willing to shell out a lot of green to get some blue. Most real estate agents will tell you that "ocean view" is the most valuable phrase in the English language (and, as you'll see, in almost every other language as well). "Consider real estate in San Francisco," said Eric Johnson of Sotheby's Realty. "There are two penthouse apartments in the same building with the same layout, but one faces the city while the other faces the water. The water view property sells for half a million dollars more."

Even though billions of people live close to water, the supply of waterfront and water view property is limited, and demand is high. Based on that combination of demand and supply and subconscious knowledge that it's good for you, one of the easiest ways we can measure and quantify the power of our ancient and newer neural maps—that is, the cognitive value of water—is by what people are willing to pay to look at and listen to.

"Happy is he who is awakened by the cool song of the stream, by a real voice of living nature. Each new day has for him the dynamic quality of birth," wrote French philosopher Gaston Bachelard.[3] Just as neuroscientists, sociologists, psychotherapists, economists, even geneticists are studying the ins and outs of human happiness, so too, ecotherapists, environmental psychologists, evolutionary biologists, and real estate agents are looking at the ways happiness and water intersect. But to begin the discussion we need to clarify the definition of happiness—a question that has intrigued and confounded humans for millennia.

A Working Definition of a Highly Subjective State

I actually detest the word happiness, which is so overused that it has become almost meaningless.

— Martin Seligman, Ph.D.

If someone asked you to define happiness, what would you say? An afternoon on the river? Success at work? A healthy and happy family? A surfboard and a perfect wave? Date night with your sweetheart? A warm puppy, as cartoonist Charles Schulz famously declared? For most of us, the definition of happiness is less a description of a feeling and more a description of the conditions that *produce* the feeling. That's because personal happiness is extremely subjective, and the final arbiter of happiness is always "whoever lives inside a person's skin," observe psychologists David Myers and Ed Diener.[4] Indeed, what we call "happiness" often contains many different emotions within it. When psychology professor and psychometrician (someone who measures knowledge, abilities, attitudes, personality traits, and education) Ryan T. Howell recently spent what he defined as a "happy" weekend in San Francisco, he described himself as feeling "energized and alive" with "engaged senses" walking along the beach; "relaxed and stress-free" while on a picnic; awe at the view from the ferry to Angel Island; a sense of community at a farmers' market; and connection with his family the entire time.[5] So was "happy" the summary of his weekend, or was each discrete experience a happy one? Or both?

Howell's experiences are a reflection of what philosophers have defined as two different kinds of happiness. The first is experienced in the moment, a result of the positive emotions that come and go based on our circumstances and internal state. Aristotle called this version of happiness *hedonia:* pleasure experienced through the senses. Hedonia is often fleeting, and so in many studies of happiness it's represented as a person's *mood*. But the sense of community Ryan Howell felt in the farmers' market is an example of the second kind of happiness, *eudaimonia:* the pleasure of living and doing well. Eudaimonia lasts beyond our moments of feeling good, and contributes to our sense of *subjective well-being*, or SWB—a common measurement of happiness used by psychologists, sociologists, and neuroscientists. SWB encompasses three key elements: positive affect (or emotions), negative affect (the number and duration of unpleasant emotions experienced), and life satisfaction (a general sense of life going well).[6] After all, it's not natural for someone to feel happy *all* the time; responding appropriately to some circumstances with negative emotions is both essential and healthy.[7] But as long as your positive emotions outweigh the negative ones and your overall life satisfaction is high—as it was for Howell during the weekend—then your SWB will be good enough for science to declare you happy.

Things get interesting, however, when we start to examine

not *how* happy Ryan Howell is, but *why*—and, by inference, what makes humans happy, both in the moment and in their lives as a whole. Why do some people seem happier than others regardless of circumstances? Cognitive biologist Ladislav Kováč once described happiness as being experienced "both as fleeting sensations and emotions, and consciously appreciated as a permanent disposition of the mind."[8] I'm sure you've known "Pollyannas," who seem to be happier than others even in difficult times; you also may have met the occasional "Eeyore," who's determined to look at the dark side of everything. Certainly we can notice and often influence the conditions that produce the "fleeting sensations and emotions" of happiness, but will that change our general disposition?

Here's where our DNA enters the picture.[9] Research psychologists Sonja Lyubomirsky, Kennon Sheldon, and David Schkade theorize that we each have a happiness "baseline" that is determined by three factors: (1) a genetically determined "set point" for happiness; (2) spending time in circumstances that make us happy; and (3) choosing happiness-generating activities and practices. They believe that while genetic predisposition accounts for around 50 percent of our level of happiness, circumstances (including the country and culture in which we live; demographics like age, gender, and ethnicity; personal history, including past disappointment or success; and life station factors like marital status, occupation, health, and socioeconomic level) contribute only 10 percent. The other 40 percent of our happiness is shaped by voluntarily pursuing personal goals through meaningful activities.[10]

On the face of it, this seems ridiculous. Why would circumstances provide so little boost to happiness? First, circumstantial happiness (as opposed to lasting joy) is fleeting and temporary. Howell enjoyed his walk along the beach, but his happiness might very well have disappeared if he'd lost his keys or injured his foot. Second, while people can control some of their circumstances (where they live, what they do, marital status), they cannot control others (past painful events or cultural conditions). If Howell had almost drowned as a child, for example, he might not enjoy the beach if he walked too close to the water. But the stronger mitigating factor for circumstantial happiness is *hedonic habituation.* "The emotional responses to a pleasant stimulus...weaken or completely cease, if a stimulus remains constant," Kováč wrote in "The Biology of Happiness."[11] The very first time Howell walked along that particular stretch of beach, his senses would have been completely engaged and alive in the novel environment. However, if he walked that same stretch of beach every day, the familiarity of the scene would decrease his engagement, and with it, the level of happiness produced by his circumstances. As a result, he would tend to fall back to the happiness "baseline" determined by his genetics. Powerful events, such as a death or repeated unemployment, can permanently shift our "happiness set point" downward, but how does it get nudged higher?

The factors that help boost Ryan Howell's happiness for the longer term are the *pursuit and attainment of personal goals* and the *adoption of meaningful activities.* Suppose that Howell was planning to participate in a race sponsored by his favorite charity, and part of his training was a daily walk on the beach. Or suppose that given his busy work schedule, spending time with his children included going to the beach every weekend. In both instances, regular beach time might contribute more to his happiness because it has additional meaning and purpose. Further, altruistic and compassionate activities that help others can boost long-term happiness. That's good news, because happiness affects our genes. In the first study of its kind, scientists at the University of North Carolina and UCLA concluded that "if we ask which type of happiness most directly opposes that molecular antipode [expression of disease-promoting genes], a functional genomic perspective favors eudaimonia."[12] In other words, while types of well-being are not mutually exclusive, psychological well-being based on a sense of higher purpose and service to others (eudaimonia) trumps consumptive pleasure (hedonia) when it comes to building better genes, thus reducing the biological markers known to promote increased inflammation linked to the development of cancer, diabetes, and cardiovascular disease. Our family connects the happiness dots by taking our kids and their friends to the beach for a day of fun on the sand and in the surf combined with an ocean-helping activity such as cleaning up trash or releasing baby sea turtles. By doing so, we are ticking several of the boxes making up "whole happiness": outside, actively in motion, pursuing goals, and solving problems together.

Activities also are more under our control than circumstances, and therefore it is easier for us to use them to boost happiness. For activities to increase our happiness, however, they must be (1) episodic and transient, thereby preventing habituation (you can't spend all day running on the beach and expect the activity to increase your happiness), (2) a good fit for the person undertaking them (if you're a vegan who rejects the killing and consumption of sentient beings, fishing might not be the best activity for you), and (3) capable of being sustained over the long term (you must be able to do the activity regularly while at the same time performing it with enough variety that you avoid habituation—changing your route for a run, for example).

The question remains, however: are certain activities and circumstances more conducive to happiness by their very nature? Is it possible that some experiences produce happiness because they reflect "universal psychological needs"—like Abraham Maslow's physiological/safety/love and belonging/esteem/self-actualization model, or economist Manfred Max-Neef's hierarchy of subsistence/protection, affection, understanding, participation, leisure, creation, identity, and freedom?[13] Or could certain circumstances—whatever our command of the situation—inevitably make us happier?

How the Brain Feels

> *Sometimes people care nothing about future or profit; sometimes, we only want to satisfy a desire, right here, right now, no matter what.*
>
> —Eduardo Salcedo-Albarán, philosopher

As Antonio R. Damasio, a pioneering neuroscientist at the University of Southern California, has stated, emotion is involved in almost every single brain activity. It is an integral part of the perception/pattern recognition/decision/action chain, often being triggered instinctively between perception and pattern recognition. "When the brain perceives a stimulus," writes Winifred Gallagher in *The Power of Place*, "whether it is birdsong in the country or the shriek of car wheels in the city, its reticular activating system, the neural switchboard for processing external and internal feedback, puts the nervous system on alert. Then it is up to us to identify how we feel about it."[14] This is a crucial distinction. According to UCLA School of Medicine clinical professor of psychiatry Daniel J. Siegel, "The rational system is the one that cares about analysis of things in the outside world, while the emotional system monitors internal state and worries whether things will be good or bad. In other words . . . rational cognition involves external events, while emotion involves your internal state."[15] This means that while our perceptions can be of internal or external events (a sensation of pain versus the movement of a potential predator), emotions *only* arise from within.

So why is this important? While neuroscience has shown that the later-developing cortical areas are involved in processing our emotions,[16] the elements of the brain that evolved first—the stem (which controls the basic life functions of breathing, heartbeat, and blood pressure, and serves as a communication hub to and from the rest of the body) and the limbic system (the "fight or flight" response area, which includes the basal ganglia, hippocampus, amygdala, hypothalamus, and pituitary gland)—are like central "switching stations" that facilitate communication between the brain and the body. Imagine that my colleague Jaimal Yogis, author of *Saltwater Buddha* and *The Fear Project*, is swimming in the ocean on a foggy day near San Francisco, where he lives. Jaimal's been a swimmer and surfer since he was a child, so he's completely at home in the water and happy to be there. His brain is merrily releasing a steady stream of feel-good chemicals: natural opiates like endorphins (creating a peaceful, euphoric feeling, also known as the "runner's high") and oxytocin (producing trust and a calm, warm mood), and the pleasure "rush" of dopamine (associated with novelty, risk and reward, exploration, and enjoyable physical activity—the same neurotransmitter that underlies many forms of addiction[17]). These neurochemicals are naturally synthesized in our bodies, a natural "medicine chest," and are released by our brains as a result of instinct and conditioned neural frameworks. Suddenly Jaimal spots a disturbance in the water's surface about fifty feet away, and his brain, which, typically risk-averse, looks for potentially negative stimuli first,[18] kicks the survival instinct into gear. Even before his conscious mind can respond, his visual cortex sends the information to his hippocampus for evaluation: is this a potential threat? The limbic system screams "YES!" and immediately the amygdala puts the body on high alert, flooding Jaimal's brain with norepinephrine (the "wake-up" chemical) and signaling Jaimal's conscious brain to "check out that disturbance NOW!" The dopamine in Jaimal's system, also triggered by the novel stimulus in a relatively calm background, helps ready the body for action. At the same time, the amygdala activates the sympathetic nervous system (SNS) to signal the rest of the body about a potential "fight or flight" situation. Jaimal looks again: is that a fin breaking through the water? Hypervigilance now becomes outright fear as Jaimal's hypothalamus (the primary regulator of the endocrine system that reacts to primal needs, like food and sex, and emotions, like terror or rage) signals the adrenal glands to release epinephrine (adrenaline) and norepinephrine, speeding up his heart rate, moving blood to the large muscle groups, and dilating the bronchioles in the lungs to provide more oxygen. All of those feel-good chemicals (dopamine, serotonin, endorphins) are overwhelmed as the stress hormone cortisol also floods the system, putting Jaimal's entire metabolism on high alert. Cortisol prompts the amygdala to continue activating the SNS while suppressing the immune response. His entire body having been neurologically hijacked by this cascade of neurochemicals, the conscious part of Jaimal's brain finally gets the message: *"potential predator—danger!"* Even though Jaimal knows very well that only one person a year dies from a shark attack in the United States, his amygdala is in high gear, laying down what are called flashbulb memories of moments of high danger[19] and screaming at him to get out of the water. He turns and swims urgently for the beach. Once there, Jaimal turns and looks back at the spot where he had been swimming, only to identify four or five dolphins' fins slicing through the water. As he stands on the sand, sides heaving, heart still pounding as the current of "fight or flight" neurochemicals starts to diminish, he mentally kicks himself for swimming away from a chance to be in the water with an entire pod of dolphins. In truth, however, he had no choice: his higher-level, cognitive brain had been hijacked by the "drive to survive" response of his emotion-based limbic system.

It can sometimes be confusing to think of all those chemicals as being linked to emotion. After all, it can seem as though the process involves little more than a mechanical series of on/off switches: see A, release chemical B; see X, release chemical Y; and on and on. And yet those chemicals fuel and form our emotions, alone, in combination, or via their absence. Some stimulate our energy (a physiological effect), but others work to make us feel happy, fearful, relaxed, tense, frustrated, focused, sad, in love—the whole range of feelings that make us truly human. Indeed, our emotions affect every decision we make, and thus who we become.[20]

Back in 1980, social psychologist R. B. Zajonc wrote, "The limbic system that controls emotional reactions was there before we evolved language and our present form of thinking. It was there before the neocortex, and it occupies a large proportion of the brain mass in lower animals. Before we evolved language and our cognitive capacities... it was the affective system alone upon which the organism relied for its adaptation. The organism's responses to the stimuli in its environment were selected according to their affective antecedents and according to their affective consequences."[21] Several decades later, that still holds. We have instinctive, emotional responses to the world around us, separate from our cognitive responses and feelings, that shape our lives and our experiences at the most fundamental level. It's easy to read a phrase like "the organism's responses to the stimuli in its environment" without realizing that "its environment" really means *the environment*, both indoors and out, internal *and* external. Furthermore, most hypothetical examples of the interaction between environment and the brain involve narratives of sudden, specific stimuli—like our account of Jaimal's encounter with what he initially took to be a shark. But not everything needs to involve the abrupt, unexpected appearance of a fin. In fact, some of the most influential signals don't come from sudden, external agitation—quite the opposite.

Today some of the same researchers who have been looking into the neurochemical and biological basis of emotions are beginning to produce findings that show exactly how our external environment interacts with and shapes our internal world. A study done in 2010, for example, demonstrated that the anterior cingulate and insula—areas of the brain associated with empathy—become more active when subjects viewed nature scenes, while in contrast, urban scenes produced greater activity in the amygdala, which is the first stop in the "danger" response and is a key player in chronic stress.[22] A second study showed that nature scenes produced greater activity in the basal ganglia, which is known to be triggered by viewing happy faces and recalling happy memories.[23] And a study done in California using fMRI showed that especially pleasant nature views activated part of the reward system in the brain—an area rich in opioid receptors that triggers feelings of wellness.[24]

The distinctions found in these studies were more refined than merely "nature versus urban"; not all natural environments were of equal potency when it came to trumping the metropolitan. The California study mentioned above found that it was coastal views that, in particular, activated the reward system. In the United Kingdom, the European Centre for Environment and Human Health is dedicated to exploring how natural water environments can help to promote human health and well-being. In 2010, researchers associated with the Centre investigated how the inclusion of water elements in natural and built scenes affected people's preferences and emotions, as well as the sense of "restorativeness" of particular environments. Subjects viewed 120 photographs of natural and built scenes, half of which had some kind of water in them, and then they were asked questions about the attractiveness of the scene, their willingness to visit the location, and how the photo made them feel. The research showed that both natural and built scenes containing water were rated more positively—with higher preferences, more positive emotions, and greater perceived restorativeness—than those without water. And interestingly, man-made structures built with water elements (houses on a canal, for example, or a plaza with a fountain) were rated as positively as green space.[25]

As Dr. Philippe Goldin of Stanford University's Center for Clinically Applied Affective Neuroscience, who examines the effects of mindfulness and meditation on human psychology, commented in 2011, "The experience of being in or beside the ocean is one filled with complex emotion based on the brain's response to environmental stimuli." The same could be said for rivers—a source of fresh water and food; lakes would also offer some of the same. It's not hard to imagine early humans acting on their innate (that is, honed over thousands of generations) emotional responses by setting up camp nearby, likely within view of, water, allowing access but far enough away to be out of harm's way due to any natural ebb, flow, flood, and meander. The resulting neurochemical release would mean that a great view that encompassed such safety and access would *feel* beautiful. Indeed, the earliest hominid records in Europe and Africa are in ancient riverine valleys. Was the view at sunrise across the water from the cave door perceived as any less beautiful to our ancestors? Perhaps even more so.

Stepping even further back in evolutionary time, we can imagine the same for our fellow primates and their predecessors and common ancestors. In fact our ability to locate, detect, sense, perceive, consume, and situate ourselves relative to water goes all the way back some 375 million years ago to the first organisms to risk a new kind of life on dry land, and begin to adapt from head to toe to the challenges of living a life as a fish out of water. If the ability to track the whereabouts of water and respond to the emotional pull of water wasn't working among these critters deep in our evolutionary past, they were dead—and dead animals don't reproduce or pass along their genes.

The Neurobiology of Happiness

. . . the brain takes its shape from what the mind rests upon.

RICK HANSON, *HARDWIRING HAPPINESS*

U.S. Supreme Court Justice Potter Stewart once famously said of pornography, "I shall not today attempt to further define [it] . . . but I know it when I see it." Until the 2000s, neuroscientists took a similar approach to the study of happiness (indeed, to the study of emotions in general), relying on study subjects' ability to know "happiness" when they felt it, and then to report it accurately. While such self-reporting is still critical to many sociological and psychological studies, for the past two decades PET and fMRI scans have given us greater insight into the neurobiology of happiness. "Such studies make it clear that human emotions are not just fuzzy feelings but 'real' in an objective scientific sense, inasmuch as they produce measurable signals in reproducible experiments," writes Oxford-based science journalist Michael Gross.[26] Today we can get a clearer picture of what happens in the brain when we happily take that walk on the beach.

However, the feeling most people identify as "happiness" actually contains many different emotional gradations within it, and therefore it triggers different neurochemicals in different parts of the brain. For example, novelty, intensity of experience, and anticipation of reward cause dopamine to flood the areas of the brain associated with arousal, motivation, pleasure, and motor control. (The desire for dopamine is also what prompts us to go the extra mile to achieve rewards.) Some of the other "neurochemicals of happiness" include gamma-aminobutyric acid (GABA), which produces feelings of calm and well-being by slowing down synaptic impulses throughout the brain; serotonin (found in the brain, digestive tract, and blood platelets), which creates feelings of tranquility, confidence, and safety; and oxytocin, the "bonding" agent that contributes to our feelings of closeness with others. Neurons communicate via impulses that together can exceed a threshold, "firing" the neuron and releasing its "action potential" to target neurons. This action potential determines which actions neurons perform (operations) and how quickly (rate), sometimes inhibiting, sometimes slowing, and sometimes speeding up these reactions in response to the circumstances. These neurochemicals affect not only the limbic system (the "emotional" brain) but also the higher cortical functions that cause us to make the decisions that determine our happiness-seeking behavior. Happiness, therefore, along with other emotions, is a brain-wide phenomenon. And in 2013 researchers reported the creation of the first "maps" of the brain activation patterns of emotion.

Situated right where the confluence of the Allegheny and Monongahela Rivers forms the Ohio River, Pittsburgh's Carnegie Mellon University is fortunate to house both an interdisciplinary department focusing on social and decision sciences and a world-class school of drama, where actors are trained in feeling specific emotions on command. Researchers exploring the "neural signatures" of emotion in the brain put actors inside fMRI scanners and asked them to demonstrate nine different emotions: anger, envy, disgust, pride, sadness, fear, lust, shame, and happiness. (Notably only two of the nine could be considered positive emotions, perhaps because the researchers were most interested in arousal and sociality, using the "calm scenario" as their baseline.) The scans revealed that each different emotion sparked varying and complex activation patterns across many areas of the brain. "Emotional experiences, like most other complex thoughts, are represented in a broad array of neural circuits," the study authors wrote. The activation patterns for different emotions were accurate enough that when researchers used a computer to compare a second set of brain scans from the same actors, it correctly identified their emotions 71 to 91 percent of the time. (Interestingly, the accuracy of identifying the pattern for happiness was higher than that of any other emotion, which suggests to me the importance of communicating happiness to others.)[27]

If happiness, along with other emotions, produces specific activation patterns in the brain, what would happen if you fired that happiness pattern over and over again? Could you turn yourself from an Eeyore into a Pollyanna? Potentially, yes—by taking advantage of the power of neuroplasticity. Our brains are wired to be Teflon for the positive and Velcro for the negative, to ensure our survival: we notice and react more strongly to negative experiences than to positive ones, because otherwise we'd lackadaisically stroll our way to extinction. However, as neuropsychologist Rick Hanson describes in *Hardwiring Happiness: The New Brain Science of Contentment, Calm, and Confidence*, if positive experiences (1) are intense enough, (2) are novel enough, (3) occur often enough, or (4) if we direct our focused attention to them long enough, they will strengthen the brain's "happiness" neural pathways and therefore make it easier for us to feel positive emotions.

How? Both the intensity and the novelty of an experience increase levels of norepinephrine and dopamine, which increase the formation of new synapses in the hippocampus and thus the creation of new neural structures. Frequent positive experiences then cause the neurons that are firing together to wire together, strengthening these positive neural pathways. And when we bring focused attention to positive experiences, we will deepen those pathways even more. "Much mental and therefore neural activity flows through the brain like ripples on a river, with no lasting effects on its channel," writes Hanson. "But intense, prolonged, or repeated mental/neural activity—especially if it is conscious—will leave an enduring imprint in neural structure, like a surging current reshaping a riverbed."[28] (Water metaphors abound in the mind/brain literature.)

Reflect on the amygdala and its primary function in reaction to stimulation—whether that stimulus is positive or negative—and helping the brain to create an appropriate response. Too much negative stimulation produces excess amounts of the stress hormone cortisol, which causes the amygdala (and therefore your brain) to become overly sensitized to the negative. Hanson points out that by consciously focusing on positive experiences and putting your full, sustained attention on them, dopamine release to the amygdala can be postponed, which will increase your sensitivity to the happiness that comes your way.[29] (Dopamine can be a "good" thing, but like all good things it's addictive, whether associated with "good" or "bad" stimuli. It may seem counterintuitive, but while happiness and empathy are essential, addiction to them is a real concern. "Empathy feels good in the moment, but it is not always the best thing in the long run," writes Loretta Graziano Breuning, professor emerita of international management at California State University.)[30]

In recent years, research has provided even more evidence of this link between the natural world and our own happiness. In a study performed in South Korea in 2009, fMRI scans of people looking at photos of natural and urban landscapes showed heightened activity in the anterior cingulate gyrus and basal ganglia (parts of the brain linked to positive outlook, emotional stability, and the recollection of happy memories) for the natural scenes.[31] But happy brains in the lab are one thing; happy people in nature are another. And "nature" is a pretty broad category, encompassing any and all ecosystems, a span of magnitudes from a town park to the ocean. Once again, our friends in Europe are in the forefront of tracking the places where people are happiest, and water settings rate highest.

"Natural" Happiness

> *Sustainable happiness is...found in our relationship with place.*
>
> — Richard Louv, *The Nature Principle*

In 2011 economics and environment researchers George MacKerron and Susana Mourato created a smartphone app called Mappiness to track the subjective well-being levels of almost 22,000 participants in specific environments in the

United Kingdom. When people downloaded the free app and agreed to participate in the study, they received signals at random times throughout the day that asked how happy they were in the moment, who they were with, where, and what they were doing. The app then used GPS to track participants' exact locations. Over 1.1 million responses—the largest number in any such study to date—were received and evaluated. The results? In general, people were happier outdoors in nature than in any kind of urban environment.[32]

While this study is notable for its large number of participants and responses, it only confirms what scientists, psychologists, and philosophers have been saying for several years, decades, and centuries, and what humans know intuitively: where we are affects how we feel. Researchers endeavoring to create a "national well-being index" examined the role of four different kinds of "capital" (human, social, built, and natural) in determining life satisfaction in communities and discovered that—surprise—"people do consider their natural environment surroundings when evaluating their life satisfaction."[33] In 2012 a pair of studies in Ottawa, Ontario, of nature relatedness and happiness found that among nearly 1,000 participants, connection with nature was "unique" in its ability to predict happiness, even after controlling for other connections (with friends and country, for example).[34] "Taking walks along the Ottawa River—fifteen minutes gives you boosts in vitality and positive emotions," Trent University psychology professor Elizabeth Nisbet, a coauthor of the study, commented at "Your Brain On Nature," a discussion hosted by Ecology Ottawa in 2013. Where we choose to spend our time matters to our sense of well-being. According to Italian psychologist Marino Bonaiuto, when there is a good "person-environment fit"—that is, our biological and psychological needs are substantially fulfilled by an environment—we experience greater happiness.[35] In 2005, Catherine O'Brien, associate professor of education at Cape Breton University in Sydney, Nova Scotia, and an expert in sustainable happiness, conducted in conjunction with the Canadian National Center for Bicycling and Walking what she called the Delightful Places Survey, to discover exactly which elements people regarded as necessary for experiencing happiness in their environment. While many survey respondents picked cities such as Vancouver, Seattle, Melbourne, and Bogotá, it was the natural environments—parks, ponds, trees, urban rivers, and waterfronts—in those cities that were considered the most delightful of all.[36]

What was it about nature that made respondents happy? Based on their answers, it was partly the immersiveness of the experience: the sounds of water, wind, birds, or even the silence; the smells of earth, water, flora, fresh air, or seaweed; the restful yet novel sights of gradations of green, blue, red, yellow, orange, or the movement of leaves and water ruffled by wind or the occasional animal or fish; the feel of cool water against a hand or foot; the yielding yet firm sensation of earth or fallen leaves underfoot; or the extra focus needed to step carefully over rock, branch, puddle, or shell. It's certainly possible that being in archetypal, recognizable landscapes (like the savannahs or shorelines our distant ancestors sought for safety and sustenance) triggers positive emotions for the adaptative reasons directly related to our survival we discussed earlier. Preliminary fMRI studies by Nik Sawe of Stanford University of the impact of natural landscapes on the brain show that in some people, the sight of inspiring natural landscapes triggers the same reward circuitry as food, sex, and money.[37]

Our happiness outdoors also could be due to active, rather than passive, engagement. Responses in the Mappiness study showed that people were happier pursuing both vigorous (sports, running, exercise, walking, hiking) and lower-energy (bird-watching, gardening, "nature watching") activities outside than they were with indoor or urban activities.[38] A major 2007 study focusing on the U.K. countryside that tracked 249 participants and summarized ten additional "green exercise" case studies found support for this correlation, concluding that walking, cycling, horse riding, fishing, and canal-boating provided significant improvement in self-esteem and mood for all demographics.[39] But how much of this happiness is due to the natural environment, and how much is due to the fact that these outdoor activities are recreational? When Mappiness researchers controlled for leisure activities, the percentage of people's increased preferences for being outdoors, in nature, remained the same.[40]

The Mappiness study showed one other interesting statistic: the highest increase in happiness in an outdoor environment occurred when people were near water. Being in marine and coastal regions added 5.2 percent to a person's level of happiness: a disparity, study authors noted, "of similar magnitude to, for instance, the difference between attending an exhibition and doing housework."[41]

Time and time again, researchers have discovered that proximity to water strengthens the positive effects that environment has upon well-being. A 2006 study of people in Ireland found that people living within five kilometers of the coast enjoyed higher life satisfaction, "other things being equal," and living within two kilometers of the coast strengthened the effect.[42] A University of Essex team that researched outdoor activities in the United Kingdom discovered that while every "green" environment had an effect on both self-esteem and mood, habitats with open water produced a significantly larger degree of improvements in mental well-being.[43] Researchers from the Institute for Hygiene and Public Health at the University of Bonn studied people who walked along promenades by the river in Cologne and Düsseldorf, and found that "water is a strong predictor of preference and positive perceptive experiences in urban environments."[44] We see this not only through the natural incursions of water into cities, but in the formation of cities themselves—particularly where we want to live.

> *Throughout history, people of all cultures have assumed that environment influences behavior. Now modern science is confirming that our actions, thoughts, and feelings are indeed shaped not just by our genes and neurochemistry, history and relationships, but also by our surroundings.*
>
> —Winifred Gallagher, *The Power of Place*

Not too long ago, most waterfront property was undesirable. It was considered dangerous, indefensible, smelly, and good only for commercial enterprises such as fishing, transport, and manufacturing. The most expensive homes were typically located in the center of town, or on the "high street," well away from the danger of marauders or of the water that residents had polluted and, in so doing, helped incubate a host of deadly (and annoying) bacteria, fungi, rot, mold, pests, and the like. For that matter, an ocean view was desirable only for seeing danger—enemies, pirates, storms, and for some, "sea monsters" perhaps—coming from afar. Today, however, in most industrialized countries waterfront property is incredibly desirable, and communities everywhere are cleaning up their rivers and old manufacturing locales and turning what was polluted, deserted stretches into chic properties, green parkland, and revitalized market space. New York City spent more than $60 million to restore the Bronx River and create parks, walking trails, and bike paths along its length. Downtown riverfronts in Portland, Chicago, Austin, Washington, D.C., and Denver are bustling. San Antonio's River Walk is the number one tourist attraction in the state of Texas.[45] Monterey's Cannery Row, so indelibly described by John Steinbeck, no longer reeks of sweat and fish guts, but instead features blocks and blocks of shops, hotels, and restaurants along the shores of Monterey Bay, right next to the renowned Monterey Bay Aquarium.

Riverbanks, beaches, and lakefronts offer a mini-course in economics, culture, and the value of environment. We've never been able to calculate the value of water other than by what people are willing to pay to live or vacation by it. But how we determine the value—experiential, monetary, or otherwise—of being by, in, around, or near the water is a critical question. With over 123 million people in the United Sates living in coastal communities in 2010,[46] what value should we be placing upon access to water and water views?

When you ask people why they like to spend time by water, the usual responses are "It feels good," "I like looking at it," and "It makes me happy." (The most common answer: "I'm not quite sure, that's a very interesting question!") When Gordon Jones, owner of Seaside Realty and Seaside Vacation

Homes, on the Outer Banks of North Carolina, surveyed thirty-five real estate agents to discover why their clients had bought oceanfront homes, the answers included:

- To hear and fall asleep to the sound of the ocean
- A status symbol
- The pull of the ocean, the serenity and respect
- The ultimate challenge of mother nature
- A good investment/rental income
- Lifestyle
- To watch the sunrises
- The convenience of having the beach at your front door
- Inspiration—people are moved to write, paint, or do whatever they do best
- To see wildlife—pelicans, whales, dolphins, sea turtles, fish

Jones explains that oceanfront lots are often referred to as being on the front row—and "who doesn't want to be on the front row?"

But what exactly *is* the water premium, and why are so many people willing to pay it? This isn't a question only for those home-shopping in Del Mar; a version of this premium is added to the bill of every water view restaurant, hotel room, or recreational facility. But because the cognitive benefits of being by the water aren't named as such on the balance sheets of many communities, economists, lacking a clear way to calculate these nonmarket values ("externalities"), often don't bother trying. Yet, by reconsidering such premiums in a different context, such qualities can become quantifiable, transformed from invisible to visible, from vaguely understandable to precisely calculated, from fuzzy ideas to cells in a decision matrix. How we think about how we think about water has begun to evolve, Blue Mind style.

What Do We Value, and Why?

> *I look out my bedroom window to the right, and I can see the ocean. And I can hear it at night. I have everything I want.*
>
> — Brett Smith, formerly homeless veteran who now lives in the Ocean Beach section of San Francisco

Scott Huettel is a professor of psychology and neuroscience at Duke University and director of its Center for Interdisciplinary Decision Science. His research focuses on understanding how the brain processes economic and social decision making—in other words, what we place economic value on, and why. "Value is one of the fundamental things your brain does," Huettel says. "In neuroeconomics, we try to understand why we value something, and how human beings trade off between the different things they value—goods, like money, and experiences, like a water view." Some value judgments are easy; choosing a new smartphone, for example, might be a matter of comparing features and ease of use among different models, and then factoring in how much you are willing to pay. It's harder to put a value on experiences, however, because the value provided is more subjective and individual, and even involves different value computations in the brain. One person will choose a hotel room with a view of the beach and palm trees, while another will opt for a room with no view and put the money toward an iPad.

Neuroscientists have determined that the ventromedial prefrontal cortex (vmPFC), located right between the eyes, at the front of the brain, is directly involved in our judgment of the value of anything. Huettel's latest research has shown that emotional engagement inspires the vmPFC to work harder (providing top-down control of behavior), and in turn we're more likely to put a greater value upon what we are evaluating. The more you care, the more you value—and overvalue. (This is why advertisers appeal to our emotions when selling their wares.)[47] But, as Huettel pointed out at the Blue Mind 2 Summit, research has shown that the value we place upon goods and experiences changes over time. In 2003, researchers Leaf Van Boven and Thomas Gilovich evaluated the value placed upon a material possession worth more than $100, or the same amount of money invested in experiences such as a trip, concert, or conversation. Overall, people rated money spent on experiences as a better financial investment and a greater contribution to their happiness.[48] Equally interesting, satisfaction with material purchases decreases over time, while satisfaction with experiences increases.[49]

If you look at the difference between the value of goods versus the value of experiences, is it any wonder that the emotional satisfaction you get from buying the iPad would diminish over time, while the emotions engendered by experiences would continue to increase as you mull over the pleasant memories they produced? (Your own life may offer many instances that prove this point—memories of childhood trips but no corresponding recollection of toys and clothes from the same period, for example.) For some people, the anticipation of a purchase or the planning of an experience—or looking at photos later on—is more enjoyable than the purchase or trip itself. And even the purchase can be secondary—online consumers often feel huge relief once they've clicked their Buy button, even if their items won't arrive for weeks.

What Will You Pay for Your Own Bit of Blue?

> *There's something really simple and idyllic about living in a house very close to the water.*
>
> —Andrea Riseborough

When you think of environments by the water, or you gaze out over an ocean or river or lake from the shore or, better yet, from a window in your home or hotel, you can feel both security and freedom, and a sense of what psychologist Marino Bonaiuto calls a good person-environment fit.[50] In *Space and Place: The Perspective of Experience*, cultural geographer Yi-Fu Tuan examined how people form attachments to homes, neighborhoods, and environments. He theorized that we feel a sense of security in the places we come to know, and freedom in having enough space to roam; we are attached to one and long for the other.[51] When we were seeking a location for our second Blue Mind Summit, in 2011, I contacted Gordon Jones to ask about booking space on the Outer Banks and to learn more about the importance of psychology in setting real estate values. For Blue Mind 2, he compared ten years' worth of Multiple Listing Service data on the price of lots located right by the ocean with those that are one, two, three, four, or five lots away from the water, and then with a group of properties that face Currituck, Albemarle, and Roanoke Sounds between the islands and the mainland. (Because the price of a house varies greatly depending upon its size and features, he felt lot sales more accurately approximated the value of a water view.) He found that the average price of an oceanfront lot was $722,608, while a lot just one row back was $420,390, or almost 42 percent less. The next most expensive lots were on the soundfront, selling for $340,747—less than 52 percent of the price of oceanfront but still more valuable than being on the third or fourth row from the ocean. Weekly rental rates for vacation homes follow the same trend. Appraisers have found that adding water—in the right place, at the right time, of course—contributes to property value more than just about anything else can, alongside square footage and lot size. In a 1993 study on Emerald Isle, North Carolina, real estate appraisers found that factors such as distance to the water, amount of water frontage, and ocean view were three of the five most important characteristics in determining value in coastal properties.[52]

Globally, water and water views impart a trillion-dollar premium on condos, houses, and all other forms of real estate. In Portland, Oregon, and Seattle, Washington, despite having no storage or basements, chic "floating homes" (formerly known as houseboats) sell for more per square foot than comparable homes on dry land.[53] Think again about that house in Del Mar. The cheaper homes behind it could have had better kitchens, more bathrooms, more room for guests, better wiring, and so on—and *it still wouldn't matter.* And it isn't just the ocean: a 2001 analysis of data for properties on the Great Lakes showed that lake views added up to an 89.9 percent premium to a house.[54] Studies of properties in the United States in 2002–2003 reveal water premiums of anywhere between 4 to 12 percent in Massachusetts, to 46 percent in Avalon, New Jersey,[55] to 147 percent for oceanfront views in Point Roberts, Washington.[56] In Singapore, an unobstructed ocean view adds 15 percent to the price of a unit in a high-rise building.[57] In 2013 in Dubai, prices for properties in the Palm Jumeirah, a development built on an artificial archipelago extending into the Persian Gulf, ranged from $580,000 for one-bedroom apartments to $5.4 million for five-bedroom villas.[58] In the United Kingdom in 2013, real estate company Knight Frank compared the price of a waterfront home with a comparable residence five miles inland and discovered that a sea view added anywhere from 26 to 66 percent to the price; views of rivers were even *more* desirable, producing an 82 percent price increase.[59] A 2000 Netherlands landscape and urban planning study discovered that water views increased the value of a property by 8 to 28 percent.[60]

Close to the water, even mobile homes can carry a hefty price tag. Paradise Cove, for example, is home to a mobile home park, right on the sand, that Hollywood celebrities call "the hippest neighborhood in Malibu," where double-wide trailers sell for $900,000 to $2.5 million.[61] Santa Cruz features another such park, right on the water, that boasts what one researcher in Pennsylvania State University's Insurance and Real Estate Department called "the world's most expensive mobile homes." "An exponentially increasing premium has typically been found, and in the past several decades these price gradients have become steeper," noted a report by the Appraisal Institute. "There are reasons to believe that will not change in the foreseeable future." World-class sunsets, waves, and a marine lab at your doorstep each morning: What premium would you pay for that? And would you regret it, just as you might regret that old toy?

"Blue" Vacations

Happiness consists of living each day as if it were the first day of your honeymoon and the last day of your vacation.

— Leo Tolstoy

Whenever traveling by jet, I look at the in-flight airline magazines and count the number of pages that have water on them. Whether it's articles or advertisements, about a quarter (sometimes as many as half) of the magazine pages fit the bill. There's a good reason: vacations by the water remain extremely popular. We're attracted to water and want to celebrate life's milestones beside it. It's no accident that there are weddings held every thirty minutes on the beach in front of one of the most popular hotels on Maui. We may be like the 46 percent of beachgoers responding to Expedia's worldwide survey who had taken a beach vacation in the previous twelve months or the two-thirds who indicated that they are either "very or somewhat likely" to take a beach vacation in the next year.[62] Or perhaps one of the more than twenty million passengers who embark on cruises from ports around the world.[63] We dream about those vacations, plan them for months (and, via neural reinforcement, expand their cognitive and emotional benefit every time we think of them), and anticipate the enjoyment and relaxation we will feel. I read once that the highest level of speargun ownership per capita is in the state of Ohio; true or not, what's certain is that people keep such equipment in the closet and dream of their next vacation in the ocean.

Social psychologist Mathew White is dark-haired, bespectacled, with an infectious grin that can warm up even the wildest, windiest days on the Cornwall coast. For the past two decades White has been studying social psychology in the United Kingdom. Now, as a lecturer at the University of Exeter Medical School and a fellow of the European Centre for Environment and Human Health, he specializes in the effects of "blue space" upon health and well-being. In 2010 White and a team of researchers examined the restorative properties of various environments. They showed forty study subjects 120 photographs of different natural and "built" (i.e., those that contained man-made structures) scenes, with different proportions of green space, water elements, and buildings. Each photo was then rated for (1) attractiveness, (2) willingness to visit, (3) perceived restorativeness (engendering feelings of being calm, relaxed, revitalized, and refreshed), (4) desire to live near the scene, and (5) willingness to pay for a hotel room with that particular view. The study subjects indicated that they felt the most positive emotions and the greatest potential for restorativeness when they looked at photographs with a preponderance of water and some green elements. (Water with some built elements was rated as restorative as natural green space without water.) Consistent with economic data, study subjects also were willing to pay more for a hotel room with views of water than of green or built spaces.[64]

Again, given our cognitive wiring, this makes sense. For most people, modern-day life produces fatigue, stress, and overstimulation, and it's important for us to vacation in places that will restore us. We seek environments that allow us to get away from our usual routines, that allow us to feel part of something greater than ourselves, to let us pursue new activities or to simply do nothing at all. For many people, that means being by the water, even if it costs a little more. Typically a "room with a view" of a beach, for example, commands a 20 percent greater price than a nonview room.[65] And because for thousands and thousands of years our neurochemicals have been prepped to release when we are staying near the water, our penny-pinching ventromedial prefrontal cortex surrenders.

The currency value of proximity to water in property is fairly easy to quantify. But the value of the economic activity in water-adjacent areas is equally, if not more, evident. In the United Kingdom, every year there are nearly 250 million visits to the coast and 180 million to other aquatic environments such as rivers, canals, and lakes.[66] According to the National Oceanic and Atmospheric Administration (NOAA), in 2011 $6.6 trillion—45 percent of the gross domestic product of the United States—was generated in counties adjacent to bodies of water. "If the nation's coastal watershed counties were considered an individual country, they would rank number three in GDP globally, behind the U.S. as a whole and China," states a 2013 NOAA report on the value of coastal communities.[67]

Yet even those figures understate the economic potency of proximity to water. As director of the Center for the Blue Economy at Monterey Institute of International Studies, Jason Scorse has studied and written extensively about the intersection of environment and economics. In 2009, he wrote, "The market value of development or resource extraction (i.e. for agriculture, timber, or minerals) is often weighed against the harder to quantify values linked to recreation, wildlife protection, tourism, aesthetics, and ecosystem services that are not priced directly in the market.... Although the prices for these goods and services are not as obvious, their values are no less real than those attributed to traditional market goods such as fish or boats."[68] Indeed, Scorse continues, the value of these nonmarket factors "may rival or even surpass the market value of the nation's ocean and coastal resources."[69]

Storms, Erosion, and Floods: The Downside of the Water Premium

Mai huli 'oe I kokua o ke kai!

— Hawaiian for "Never turn your back on the ocean"

In 2012, I volunteered to help clean up Coney Island a few days after Superstorm Sandy. I grew up going to Coney Island, and the devastation, the mangled amusement park, the mounded beach, clashed with my vivid childhood memories of a fantasyland by the sea. That fall day the water was cold but the sun was out, and people were walking on the boardwalk with their shirts off. A few days ago the ocean had been Public Enemy No. 1; now folks were at the business of cleaning up and working on getting the electricity back.

I stopped by a gift shop called Lola's, filled with the kind of cool, kitschy, glittery stuff that defines popular summer beaches—I was the first person after the storm to walk into her dark, wrecked, powerless store. I met the owner, Lola Star, who, besides selling me a T-shirt for my daughter, told me that before Sandy hit she had moved all of her inventory from the boardwalk location to her boutique in the Stillwell Avenue subway station because she assumed it would be safer there. But the subways flooded, and Lola lost $25,000 worth of that inventory. It would take another $25,000 to rebuild, and she lost $50,000 in sales while the shop was closed. But seven months later, she reopened her store in the subway station. Her customers had helped her rebuild. Now, cleaned up and put back together, Lola was in business again.

"The people here need me," she said as we looked across the boardwalk, the beach, out to the glistening Atlantic Ocean. "They need love, they need Shimmer." (Shimmer is her dog.)

Over the past decade we've seen too many instances of the dangerous side of the water premium. The tidal surge and flooding of Sandy in 2012 and Katrina in 2005 . . . the tsunamis in Japan in 2011 and in Indonesia, Thailand, and Sri Lanka in 2004 . . . the floods along the Mississippi in 2002, 2008, and 2012 . . . the great European floods of 2013, which caused damage costing more than $16 billion, the area's costliest natural disaster[70] . . . the 2014 floods in England . . . and that doesn't take into account the slow, steady, corrosive effects of natural erosion and water spray upon wood and stone. The water giveth, and the water taketh away.

Of course, you can't stop the water from undermining the structures we are determined to build next to it.[71] But you also can't easily stop the people who want to live by it and enjoy it, because you can't short-circuit their hardwiring. In 2013, Mathew White and his colleagues used data drawn from a 2009–2011 Natural England study of 4,225 survey respondents to discover that, compared with visits to parks and open spaces in towns and cities or visits to the countryside, visits to coastal areas produced significantly greater recalled feelings of restoration across all demographics.[72] It's hard to override sympathies that have built up over hundreds of thousands of years, even when warned that your livelihood, if not your life, could be in danger should the waves turn to teeth.[73]

In the summer of 2010 I traveled to the scene of another disaster, this one man-made. The BP oil spill spewed more than 210 million gallons of crude oil into the Gulf of Mexico for eighty-seven days, fouling 16,000 miles of coastline in Texas, Louisiana, Mississippi, Alabama, and Florida.[74] As I helped to survey, rescue, and clean turtles and other wildlife coated with oil, I thought also about both the economic and psychological costs of the disaster. The spill had decimated an area that provided 30 percent of U.S. oil production and 20 percent of natural gas at the time, and 25 percent of the nation's seafood. Tourism in the entire Gulf Coast area was at a standstill. And the environmental costs were, and are, incalculable.[75] People in these communities not only had their homes and their jobs destroyed, but also their mental health. Researchers at the University of Maryland found that even those in the area who had not been directly affected by the spill were feeling more anxiety, depression, and hopelessness—a kind of PTSD.[76]

But the costs of the BP oil spill pale in comparison to the damage caused when water simply does what comes naturally—flooding the areas along rivers in spring, washing beach sand out to sea in some places and replenishing it in others. We have interfered with the natural actions of water by trucking in sand to preserve our beautiful beaches, or changing the course of rivers so we can navigate them more easily and build more homes and businesses along their banks. But, as Mark Twain described efforts to reroute the Mississippi back in 1883, "Ten thousand River Commissions... cannot tame that lawless stream... cannot save a shore which it has sentenced; cannot bar its path with an obstruction which it will not tear down, dance over, and laugh at."[77] And all too often we pay an increased price for our attempts to turn water from its natural path or action. Flood insurance costs in the United States are skyrocketing, and because of the losses they suffered from events like Hurricane Katrina and Superstorm Sandy, insurance companies are increasing premiums for anyone who lives near water. Worse yet, as the climate warms and oceans rise, the water is taking back its own. Chunks of cities such as Miami, New Orleans, even parts of New York City, could be underwater in the next few decades. Marshlands all over the Gulf Coast are disappearing. And severe weather events that produce hurricanes and floods are becoming more regular.

Atop many of the houses in Nantucket and Martha's Vineyard you will see a railed rooftop platform popularly called a widow's walk, designed so the wives of captains and sailors could tread back and forth, watching and waiting for their husbands to come home. As the name romantically suggests, all too often they watched in vain. They feared those who went to the sea would be consumed by it. Today, we fear the sea will consume us.

That we've made this situation more dreadful by not taking care of our planet, and by our own reckless inclinations, actually validates the positive power of Blue Mind, because Blue Mind is all about recognizing the creative disequilibrium (and occasional balance) within a larger whole. Not only can getting your Blue Mind on make your life better, it can make the lives of others better. You can't hold back the water—but you can swim with the current.

4

The Senses, the Body, and "Big Blue"

It is only through the senses that we experience what it means to be fully human.

— Laura Sewall, *Sight and Sensibility*

As we've seen, previous experiences shape the way our brain reacts to the world. But it's more complicated than that, and to better understand the power of Blue Mind, we need to look more deeply at how the brain takes in information and transforms it into predilection or revulsion.

Scientists believe a baby's senses begin to develop as early as eight weeks after conception. By twenty-six weeks, all five of the "traditional" senses—touch, taste, smell, hearing, and sight—are present and functioning even in the womb.[1] But after birth our senses start to engage with the wider world, and we are quickly awash in a sea of perceptions, with the brain as the processor that turns a sensation into a perceptive experience. It accomplishes this remarkable feat in several different ways. First, the brain breaks down this onslaught of sensory information into more manageable "bytes." A byte is a unit of digital information consisting of eight bits, each bit designated by a 0 or a 1. Comparisons made between the human brain and computers are problematic in many ways—your brain is so very much more complex than a computer, at least in the way computers most people know, use, and love today function—but do raise the question: what *is* the storage capacity of the average brain? Estimates vary considerably, ranging from as low as 1 terabyte (10,000,000,000,000 bits) to around 2.5 petabytes, 1 petabyte being around 1,000 terabytes.[2] Regardless of capacity, there are limits to how much and how fast we can take up information; David Poeppel describes the brain as having a "sampling rate" (he's speaking about visual and auditory stimuli, but the principle holds for any kind of perception). "The world comes at you, visually or auditorily or through your other senses, and the brain's solution is to break the information down into small packets, which it takes in at particular rates of time," he explains. "The job of the brain is to put all of those 'packets' of information on top of each other, integrate them on the fly, and make them congruent." The body's systems for transmitting data from the senses to the brain also limit the amount of information coming in at any given time. Even though upwards of 10,000,000,000 bits of visual stimuli hit the retina every second, due to the limited number of outputs from the eye into the optic nerve only 6,000,000 of these bits can be transmitted from the retina through the optic nerve, and only around 10,000 bits arrive at the visual cortex. Of those 10,000 bits, around 100 are actually used for our conscious perception of what we are seeing.[3] Yet here the brain's fantastic strength as a pattern recognizer allows it to fill in missing information that your senses can't perceive; for example, you never notice the "blind spots" in your field of vision in both your left and right eye because your brain automatically fills in the necessary visual data.

Second, the brain separates the "signal" from the "noise," so to speak: there's a specific location in the brain called the inferior parietal lobule (IPL), at the intersection of the occipital (vision), temporal (hearing), and parietal (touch) lobes, that's designed to receive, process, and integrate sensory data in such a way that the world makes sense. At any given moment you may be seeing, hearing, tasting, smelling a huge number of things, but your cognitive mind uses only the stimuli needed to turn that information into a cogent, multisensory experience.[4] At the same time, too much paring down is a cognitive malfunction of the highest order. If we try to isolate the input from our different senses, we lose much of our ability to find meaning in the world.[5]

Third, the brain uses past experiences to build perceptual "maps" (neural signatures, especially corresponding to dynamic sensory inputs as opposed to our hardwired static networks, are often described in cartographic terms), and then uses those maps it has created to make sense of new perceptions.[6] We have maps that allow us to discern shapes and see colors, pick out the voice of a loved one, smell whether milk has gone sour, or determine whether the temperature of our bathwater is just right. In fact, we may well base our own version of the "real world" not solely on what we perceive but instead by matching what we perceive with what we think we *should* perceive. "Perception is based on recognition and interpretation of sensory stimuli," writes neurologist Gerald Smallberg. "From this data, the brain creates analogues and models that simulate tangible, concrete objects in the real world.

Experience...colors and influences all of our perceptions by anticipating and predicting everything we encounter and meet."[7]

If we don't have perceptual maps for a particular set of stimuli, the brain has a great deal of difficulty making sense of what it perceives. Studies of individuals who were either blind from birth or lost their sight very young and then had it restored years later show that while they immediately could discern light and color,[8] the previously blind typically had difficulty recognizing shapes, dimension, closeness or distance of objects, or understanding the meaning of facial expressions in others. In this situation our old acquaintance neuroplasticity is both a friend and a foe. If someone has had little or no visual stimulation growing up, usually the other senses (hearing, touch, smell) will "take over" parts of the brain normally devoted to visual processing. When visual stimulation is restored (or a substitution is made), the brain immediately begins laying down neural networks to support vision—but depending on how much time has passed, the spaces usually reserved for vision are now occupied, forcing those new networks to be laid down elsewhere. On the flip side, because of neuroplasticity it is possible for the blind to learn to "see," sometimes by utilizing some very unconventional methods. David Eagleman describes a device developed by neuroscientist Paul Bach-y-Rita in the 1960s. A video camera was mounted on the forehead of a blind person, and the picture of the environment was translated into small vibrations on the back. After a week, the blind individual was able to "see" the environment through his back. In a more recent case, a rock climber learned to "see" through a grid of 600 electrodes placed on his tongue.[9] Humans have even learned how to coarsely echolocate and "see" like marine mammals or bats, by listening carefully to the way clicking sounds they made with their mouths or walking sticks bounce off surrounding objects, buildings, and landscapes. This seems impossible to the rest of us—and it *is* impossible for us, because our brains are populated with standard systems, with no room for such radical remodeling.

Like the neural networks we lay down and strengthen by what we pay attention to and the actions we take, we build our perceptual maps based on the sensory experiences we pay attention to in the world. That's why artists and photographers are usually far more precise about color, line, and sharpness of visual images: they have spent years developing perceptual maps for visual distinctions. "Our attentional focus, both internally and externally, influences and creates subjective reality by facilitating the perception of some objects, relations, and events to the exclusion of others," writes ecopsychology pioneer Laura Sewall. [10] At the same time, it turns out that such exclusion can have surprising, sometimes hazardous, spillover effects. According to Michael Merzenich, "One of the more important findings of recent research is how closely connected our senses (hearing, vision, and so on) are to our memory and cognition. Because of their interdependence, a weakness in one is often related to—or even the cause of—a weakness in the other. For example, we all know that Alzheimer's patients slowly lose their memories. One way this manifests is that they eat less food. Why? As it turns out, visual deficits are also a part of Alzheimer's. People eat less because they can't see the food as well.... Another example is in normal age-related cognitive changes. As we grow older, we get more forgetful and distracted in large part because our brain does not process what we hear, see, and feel as well as it once did. The result is that we can't store images of our experiences as clearly, and so have trouble using them and recalling them later."[11]

But it isn't just our perceptions of the physical world that shape us, and that are in turn shaped by what we perceive. Our emotional reactions to what we perceive are also encoded in our brain as part of the "formula." The problem is that we have lost sight of this important fact.

> *I go to nature to be soothed and healed, and to have my senses put in order.*
>
> —John Burroughs

Imagine yourself walking in the forest along the edges of a stream. The green canopy of leaves overhead rustles slightly, moved by wind or perhaps the passage of a bird or squirrel through branches. The uneven ground beneath your feet makes you pay more attention than usual to the feel of the earth, the roughness of tree roots, the occasional rock that juts upward to trip you if you're not careful. You smell the slightly fermented odor of damp leaves rotting, and you get an occasional whiff of spray from the water beside you. You notice the roughness of the bark on the trees you're passing, and the gradations of green of various trees and bushes. Your ears, without the usual electronic and/or urban overstimulation, start to hear the different ways your shoes sound on dirt or on leaves. Unconsciously, you had been aware of the sound of the water in its streambed and a raft of birdcalls, some right overhead, many more in the distance; but as you walk, these sounds come into focus. You stop for a moment and simply enjoy your surroundings. Your senses are doing what they were meant to do: engage fully with the natural world as an integral part of it. "For the largest part of our species' existence, humans have negotiated relationships with every aspect of the sensuous surroundings," philosopher and cultural ecologist David Abram states. "And from all of these relationships our collective sensibilities were nourished. Direct sensuous reality...remains the sole solid touchstone for an experiential world now inundated with electronically-generated vistas and engineered pleasures; only in regular contact with the tangible ground and sky can we learn how to orient and to navigate in the multiple dimensions that now claim us."[12]

Yet in most of the modern world, our primary perceptions are almost entirely filtered through human construction. We walk on sidewalks or drive on roads in our human-built cars, to places of work that are human-created, "built" environments. We listen to music, watch TV, surf the Internet, read books; eat prepared meals that taste nothing like their original natural ingredients; use perfumes and soaps and household products that mask real smells; and touch screens and plastic and processed materials that feel like nothing in nature. None of this is bad *per se*, but we have been separated from the very stimuli we were built over hundreds and hundreds of thousands of years to perceive: the richness of the natural world. Neon signs are beautiful, cell phones are useful, subways are efficient, and crowds spilling out into the paved streets have a powerful appeal. But such commotions are the wrong key to a lock installed by evolution—and even those amidst the skyscrapers know it without knowing it. Study after study, as well as personal experience, shows that the overstressed, overstimulated, urbanized mind can find greater relief in the more subtle perceptions of a park, a forest, a beach, or a riverbank than it can from almost any human-produced environment.[13] And while visualizing yourself in natural environments is certainly better than nothing, all of our senses are craving the full "nature" experience.

Sight and Water

> *It is life, I think, to watch the water. A man can learn so many things.*
>
> — Nicholas Sparks

According to neuroscientist V. S. Ramachandran, there are as many as thirty visual areas in the brain devoted to everything from identifying what something is, to where it is in the environment, its relationship to other objects, as well as relationships between features of the object itself. Other associated areas provide a name for the object as well as all the facts and memories associated with it (in other words, creating its meaning), while the amygdala assesses its emotional significance.[14]

Many studies have been done that show exactly what makes certain visual images stand out. Not surprisingly, these characteristics are tied to what helped us to survive in the natural world and are prominent features of water: color, shininess, and motion. Let's start with color.

Color

I hope the Pacific is as blue as it has been in my dreams. I hope.

— Stephen King, from "Rita Hayworth and Shawshank Redemption"

In Harpur Hill, near Buxton, Derbyshire, British locals know a quarry pool as the Blue Lagoon. For years it's attracted dozens of swimmers to its beautiful turquoise water—despite the signs posted at its edges saying things like "Warning! Swimming in this pool can cause skin and eye irritations, stomach problems, fungal infections, and rashes." The pH level of the quarry pool is close to that of ammonia or bleach, and the water is filled with rubbish and dead animals. Yet it wasn't until the local government dyed the water in the Blue Lagoon black that swimmers stopped taking the plunge.[15]

Whether it's logical or not, humans seem drawn to the color blue. It's overwhelmingly chosen as their favorite color by people around the world, beating its closest competing color by a factor of three or four.[16] Both women and men prefer blue to green, red, or purple.[17] And blue is everywhere: while it's the rarest color in nature (appearing only occasionally in plants or animals), on every sunny day we can look up and be dazzled by the incredible blue of the sky. Water, too, possesses a range of blue tones, mixed in with greens and browns and whites depending on depth and location. And, as the residents of Harpur Hill demonstrate, even toxic blue water can entice us with its sense of cool refreshment.

When marketers and psychologists have asked people

what qualities they associate to blue, they use words like "credibility," "calming," "clean," "focused," "cleanliness," "openness," "depth," and "wisdom."[18] Emotionally, blue is associated with trust, confidence, and dependable strength: is it any wonder that companies such as Facebook, AT&T, Lowe's, American Express, HP, IBM, Walmart, Pfizer, and Vimeo use blue in their corporate logos? (Blue even predominates on the packaging of black-and-white Oreo cookies.) You can open almost any magazine and see several ads that use beautiful washes of blue to market everything from tropical or ski vacations to the latest offer from Best Buy or Bed Bath & Beyond. Even those who are well aware that the success of their products involves reducing a sense of calm and increasing a sense of urgency go blue; witness the logos of Facebook and Twitter.

Other investigations have confirmed this soothing effect. For example, researchers in Japan reported that people who sat next to a blue partition while playing a video game had a more regular heartbeat and reported that they felt less fatigued and claustrophobic than those who sat next to a red or yellow partition.[19] And in a recent "virtual reality" study, subjects who were wearing bands on their wrists that transmitted a sensation of heat were told to signal when the temperature reached a painful level. As they did so, they watched monitors that showed a picture of their arms with the band represented by red, green, or blue. People perceived the greatest pain when the area was colored red, and the least pain when it was colored blue.[20]

Why make so much of fragmentary blue
In here and there a bird, or butterfly,

Or flower, or wearing-stone, or open eye,
When heaven presents in sheets the solid hue?

Since earth is earth, perhaps, not heaven (as yet)—
Though some savants make earth include the sky;
And blue so far above us comes so high,
It only gives our wish for blue a whet.

—Robert Frost, "Fragmentary Blue"

Light is an electromagnetic wave—not metaphorically but literally—and color is a function of the length of the wave in each packet of light that enters your eye after it reflects and refracts off and through a variety of surfaces and substances, from air and water, to skin, fur, and feathers. Violet is at one end of the visible color spectrum, which ranges from 400 to 700 nanometers (nm), and red at the other, with blue at about 475 nm. Exposure to light of the wavelength corresponding to the color we call blue has been shown to produce physical, cognitive, and emotional benefits. As Amir Vokshoor, a neurosurgeon who specializes in minimally invasive microsurgical treatment of adult spinal and cranial disorders, remarks, "Due to its specific wavelength, the color blue is known to exert a calming, relaxing, yet energizing effect and thus stimulate a positive emotional response. In fact, the arousal mechanism stimulated by blue's wavelengths correlates to the release of neurotransmitters thought to be associated with feelings of euphoria, joy, reward, and wellness related to the effects of dopamine." Vokshoor theorizes that the reason blue produces such positive feelings is that humans evolved on a planet that is primarily shades of water and sky blue, alongside shades of green and earth tones.

In a 2010 study, seventeen people listened to voices while they were exposed to alternating periods of blue or green light. According to fMRI scans, blue light enhanced the connectivity between the area of the brain that processes voices and the amygdala and hypothalamus (the body's primary gateways of emotional processing).[21] That is, blue light actually strengthened and further established the crucial neural networks that allow us to hear and understand voices. Color clearly colors our other senses, and vice versa.

It's far more fun to push past "blue is good" and ask—at the risk of offering up a just-so story—"why would we have evolved for it to be so?" From an evolutionary point of view we can imagine that auditory function being enhanced in response to the color blue could indicate proximity to open sky and water, implying a need to listen more carefully for distant sounds than when in a more closed or confined locale. Or, while we are conjecturing, couldn't it have been because the "white noise" of waves and current made it harder to hear, versus the quieter savannah?

It isn't just the colors that we see which have a pronounced effect on how we understand the world—it's the very means by which we see those colors (and everything else) in the first place. Humans need the full spectrum of light available in nature to be healthy and to regulate our circadian rhythms. Unfortunately, for reasons related to production cost and the ubiquity of cheap energy, most incandescent light sources concentrate on the yellow-orange-red end (similar to firelight) and lack the blue-green end of the light spectrum. Moreover, the modern lifestyle that keeps us up late and gets us up early can limit our exposure to natural light. That's why exposure to blue-light wavelengths has been shown to readjust circadian rhythms in nighttime workers and lessen the effects of seasonal affective disorder (SAD). The future of lightbulbs lies in full-spectrum, adjustable, and customizable LEDs, which are good for illumination and, thanks to their range of spectrum, good for your brain, too.

The LED point isn't that flicking a switch instantly lights up your Blue Mind, too. You still need the water. But it further reinforces the neurological potency of that most central color when it comes to waves, currents, pools, and the like.

Of course, as anyone who's ever had the "blues" can attest, blue has its dark side. It can be associated with coldness, sorrow, even death (perhaps because our lips and faces take on a blue pallor when we are low on energy or oxygen).[22] In 1901 Pablo Picasso sank into a profound depression and began what is known as his Blue Period, when he used mostly blue and blue-green hues in his paintings. Muddy Waters describes the blues as "deep, profound, with a primordial groove." And yet interestingly, when depressed people are asked to describe the color of their emotions, gray is the word more frequently used than blue—and they often still describe blue as their favorite color.[23]

Indeed, at a fundamental level our brains can't let go of blue's powerfully positive qualities. In Japan, suicide is a significant problem and jumping in front of trains a common method. Several years ago the government installed blue lights in high-crime areas as well as railway stations. Crime dropped by 9 percent, but more important, suicide attempts in the blue-lit areas ceased altogether.[24]

Such a decrease may be due to blue's calming effects—or perhaps its ability to enhance cognition. According to a 2010 study in Europe, exposure to blue light increased responses to emotional stimulation in both the hypothalamus and the amygdala—areas of the brain that involve attention and memory.[25] Recently two researchers in Canada showed that red and blue enhance different kinds of cognition. Red seemed to be helpful for attention to detail, for practicality, and for specifics, whereas blue encouraged greater creativity and attention to relationships between items. "Depending on the nature of the task, different colors might be beneficial," the researchers commented.[26] Possible crime victims and perpetrators alike might behave differently if they were thinking more clearly; likewise potential suicides. Criminals might even contemplate the odds of getting caught and decide to back off.

"The color of young dreams is changing from green to blue."
— ERIC CHARLESWORTH[27]

One of my favorite pieces of art glows with blue: the Chagall *America Windows* at the Art Institute of Chicago. Over the years a hazy gray film accumulated on the glass, but not too long ago the windows were restored and the blue—"Chagall blue," as it's called by lovers of his work—is vivid again. The blue illuminates everything it touches, bringing coolness, freshness, and clear light to the space. Whenever I have a chance, I visit the museum. There I stop and gaze at the depth of color ranging across Chagall's three windows, 8 feet by 32 feet of watery immersion. When I stand in front of them, I think, *If anyone doubts the power of blue, try painting the bottom of a swimming pool red!*

Motion Without Movement

Today, almost no pool goes in without a water feature. They want to see the water move; they want to hear it move, they want it to dance, to turn colors at night.

LYNN SHERR, *SWIM: WHY WE LOVE THE WATER*

What is it about the way water moves, reflects, glimmers and glows that mesmerizes and transfixes us? The reflective surface of water draws humans in the same way it drew our ancient ancestors to drinking pools in Africa—after all, the shiniest thing our ancestors probably saw was sunlit water. Today, it seems that our attraction to the sparkle of watery surfaces is part of our DNA. A recent study of children as young as six to seventeen months demonstrated their universal inclination to lick the shiny surface of a stainless steel plate (or the mirrored surface of a toy) while on their hands and knees—"in a manner not unlike the way older children drink from rain pools in developing countries," stated the researchers. "Such mouthing of glistening surfaces by nursing-age children might characterize the precocious ability to recognize the glossy and sparkling features of water long before this information is useful later in development."[28]

"Near water, light takes on a new tonality," writes French philosopher Gaston Bachelard. "It seems that light has more clarity when it meets clear water."[29] It's why humans love fountains and waterfalls, why we are transfixed by sunlight sparkling on the surface of ponds, lakes, streams, rivers, and oceans. In the motion of the water we see patterns that never exactly repeat themselves yet have a restful similarity to them.

Our eyes are drawn to the combination of novelty and repetition, the necessary criteria for the restfulness of "involuntary attention." Water becomes something that people can watch for hours and never be bored. Indeed, images of water in motion can even modify a negative response to other environmental stimuli: a 1999 study showed that white noise was considered less annoying when it was accompanied by a picture of a waterfall, suggesting a natural sound source.[30]

Even more healing, perhaps, is the lack of speed with which the visual stimulus of water's movement enters our eyes and brain—all the more potent in a world of rapid, digital, flicking visual images in our media; commercials, action sequences in movies and on TV, and video games coming at us at a pace that's too fast for the brain to comprehend consciously, putting us into a state of hypervigilant overstimulation that all too quickly becomes our status quo.[31]

But if we stop for just a moment, put our tablets and smartphones down, and just watch the water, we can find a measure of rest and a sense of peace in the dance of drops from a fountain or the lazy motion of the current in a broad river. As reporter Charles Fishman writes in *The Big Thirst,* a book about the future of water as our most precious resource, "It's hard to be in a bad mood around beautiful flowing water. Whatever cares you have are lightened when you spend some time with water. The presence of a brisk, bright mountain stream makes you smile, it makes you feel better, whether you're already feeling good or you're low."[32] And that ability to sit outside and watch the play of water has benefits beyond contentment. Ophthalmology studies done in Australia theorize that the increasing rate of nearsightedness in children and young people could be due to less time spent in settings where the eyes need to focus on longer distances.[33] In an age of computer/smartphone/screen-stressed vision, the homonyms "sea" and "see" take on a significance beyond rhyme.

The Redolence of Water: Smell and Taste

> *I believed I could identify the scent of the sky as I stood there, a blue menthol fragrance similar to the scent of seawater that sprayed into my face when I first dove into the ocean. That initial scent was much more subtle than the ocean's heavy, fishy aroma; it was a whiff of salt and mint, just as I approached the water on a dive, that warned me that a more powerful scent would soon enter my nose.*
>
> —ANNE SPOLLEN, *THE SHAPE OF WATER*

Certainly we don't want the water we drink to have much of a smell (or taste),[34] but if you think of any moment in which you encountered water, chances are you will remember some aroma, even a subtle one. The smell of rain in the air, or the cool mist of droplets from a fountain . . . the slightly woody scent of leaves mixed with freshness when you come upon a creek or stream . . . the tang of salt along the ocean or the just-can't-quite-put-my-finger-on-it fresh smell of a lake. All of these scents have natural origins—plant decay, bacteria, dimethyl sulfide (DMS), ozone.[35] These bits and pieces of life and death, animal, plant, and mineral, produce a raft of aromatic chemicals that enter the nostrils, triggering chemoreceptors that stimulate the olfactory receptor neurons, which send signals to the brain. Unlike our other senses, however, information about smell bypasses the thalamus and goes straight to the olfactory bulb, a part of the limbic system that includes the amygdala (which, as we've discussed, is crucial in both the formation and memory of emotional experiences). This is the reason that smells are some of our most powerful emotional triggers.

Numerous studies have shown that smell may affect our cognition, mood, and health. In a 2011 paper, psychologist Andrew Johnson states, "Essential oils and other commercially available scents have . . . been shown to positively affect memory, vigilance, pain perception, self-perception/confidence, consumer decision making, and alertness."[36] An earlier study showed that using orange and lavender oils on patients prior to treatment reduced their anxiety and improved their mood.[37]

Unfortunately, there have been no studies of the effects of more watery scents on humans, but I believe that the strong emotions many of us experience around water can be triggered by even the faintest whiff of salty air or the fragrance of damp earth after rain. Not long ago, I asked master perfumer Libby Patterson to create an ocean scent for me. She did, and calls it Wave. It's a mix of oils from the coastal vegetation near my home, the smell of kelp, mixed with burnt shell. To me it hints of the way my skin smells in the morning after swimming in the Pacific Ocean all day and sleeping by a fire on a beach all night. I now travel everywhere with a small vial of it in my kit, to remind me instantly of home on the Slow Coast.

But even more indefinable than the smell of water is its taste. Taste and smell are intricately entwined, the one literally "feeding" off the other;[38] but where smell hits the brain directly in its limbic system, to taste something the brain must use the senses of taste, smell, touch, texture, and heat (and even, sometimes, pain), processing the information in the gustatory area of the cerebral cortex. Many people have strong likes and dislikes of certain tastes that may have been instilled before birth. (What your mother ate when you were in the womb can have an effect on your preferences.) Some of us prefer sweet tastes, others salt, others bitter, and so on; certain people can't stand particular textures, or the smell of a food (Brussels sprouts or broccoli) puts them off, or they can't take even the smallest amount of heat/spice. On the other hand, most people have certain foods that trigger strong emotions. Your mom's apple pie or baklava, the mac and cheese you had every Thursday night when you were a kid, your grandmother's dal or *kofte* or collards or pierogi—even the smallest taste will call up the memories of your childhood.

For many of us, the taste of water is the taste of salt—or even more, the taste of the creatures that make the water their home. Who hasn't tasted a freshly caught piece of fish, or lifted a succulent mussel, clam, or oyster to their lips, and thought, "This tastes like the sea or lake or river"? And if you've ever had the chance to eat that piece of seafood close to the water, you'll know how much your gustatory enjoyment can be enhanced by the rest of your senses. In 2006 experimental psychologists Charles Spence and Maya U. Shankar conspired with celebrity chef Heston Blumenthal to conduct an experiment to see how surroundings affect taste. At the opening session of a conference in the United Kingdom on art and the senses, guests were served oysters with two different soundtracks in the background: one a "sounds of the sea" complete with seagulls and waves, the other a series of farmyard noises that included chickens clucking. Not surprisingly, the audience rated the oysters consumed to the sounds of the sea as more pleasant.[39] After this experiment, Heston created a "Sound of the Sea" tasting menu for his restaurant The Fat Duck, in nearby Bray. The plate resembled a beach, with sand, seaweed, and foam decorating it, and a pair of iPod headphones peeking out of a seashell surrounded by food. When guests put on the headphones, they heard a "sounds of the sea" soundtrack to accompany their meal. See, hear, smell, taste—is it any surprise that the dish became a Fat Duck signature?[40]

Heston's meal points out the importance of something we haven't focused on yet: sound. We'll get there soon, but first let's get to another sense: touch.

The "Feel" of Water: Body, Brain, and Your Senses

> *It appeared as if I had invited the audience into the water with me, and it conveyed the sensation that being in there was absolutely delicious.*
>
> — Esther Williams

Stand next to a rushing waterfall and you'll feel the vibrations in your bones. Drift along on a boat in a calm lake and notice how the gentle motion makes you sleepy. Swim laps in a pool and revel in the fact that your body is light, buoyant, supported completely by the water that surrounds you. Stand in a hot shower at the beginning or end of a day and enjoy the tension draining from your muscles. The body provides a cornucopia of different sensory stimuli to the brain. These sensations anchor us in the world by giving us a clear indication of our physical position in it, and bring us into the present moment when our thoughts wander. Without the senses that reside in the body—touch, pressure, temperature, weight, motion, position, balance, vibration, and pain—we could not interact with our outer environment safely.

Skin, bones, muscles, joints, and organs—all transmit information from their nerve cell receptors through the sensory nerves in the spinal cord, into the thalamus (also the primary nexus for sight and hearing) and then to the different areas of the brain depending on the particular sensory input. Touch is processed in what is known as the somatosensory area, with more neurons dedicated to signals from the sensitive parts of the body, such as the face or hands than, say, the shoulder. Pain, proprioception, and posture are processed in the cerebellum; there are also pathways for the kind of perception that maintains muscle tone and posture without your conscious thought. To understand that, try consciously letting your body sag and go limp all over right now. That's probably not your typical stance, even when you're at ease and not paying any particular attention to your posture. Additionally and simultaneously you maintain balance with the vestibular system, which runs from the inner ear canals to the brain stem, and then to the cerebellum and the reticular formation.

As we've already seen when discussing neuroplasticity, our bodies are ever-changing within an environmental feedback loop. Thus, our senses help us perceive the outside world, but that outside world can deeply inform our senses themselves, which in turn can stimulate our attention on the outside world. Among the foundational elements,[41] water can far and away play the most beneficial role in keeping us "in touch" with the sensory world. It has a tangible quality, a weight, heavier than air yet, unlike earth, we can move through it.

Like earth, water supports us and takes our weight; in fact, because the human body has a density similar to water's, we are buoyed up by it. We feel as if we weigh less in water, which makes it the ideal medium for exercise for those with physical limitations. "The body, immersed, feels amplified, heavier and lighter at the same time. Weightless yet stronger," writes swimmer Leanne Shapton.[42] We couldn't have survived without fire, but the centrality of water is unquestionable.

Our senses keep us firmly connected with the world around us—yet at the same time, they trap us in what evolutionary biologist Scott Sampson calls "our skin-encapsulated selves" on the online salon Edge.org (in response to the question "What scientific concept would improve everybody's cognitive toolkit?").[43] It's easy to forget that the body is composed of the same atoms that make up the world that surrounds it, and that we are exchanging molecules with every inhale and exhale, every morsel of food, and every skin cell that sloughs off as a new one appears. By cutting ourselves off from the awareness of our interdependence with everything that is not man-made, we sever ourselves from the incredible beauty of what our senses actually are taking in. Yet when the architectural beauty of the tower, bridge, street, cafe, tunnel, walkway, taste, sound, and smell of the village, town, city, and megalopolis are woven and suffused with blue and green spaces, life can be so very good.

Flotation Tanks: The Brain on Nothing

"Something is happening," your body says to your brain, with mild urgency. My brain went a little haywire. When the storm

passed, I found myself in a new and unfamiliar state of mind.... For the first time in my waking life, I had zero thoughts. It was a mental quietude I'd never known existed.

— Seth Stevenson

Near water, but especially in water, our bodily senses—touch, pressure, temperature, motion, position, balance, weight, vibration—are truly alive. But there is one place where being in water can feel like nothing at all; where, very deliberately, none of our senses are stimulated, with some very unusual results for the brain. In 2010 sound artist and musician Halsey Burgund and I collaborated on a project called Ocean Voices. We asked people of all ages from all over the world to talk about water, the ocean, and their feelings, and recorded their voices to create a symphony of ideas expressed through the spoken word. Many of them recalled the physical sensations of being in water.

"I could relax and float."
"I feel weightless."
"It's like floating in infinity."
"I feel the up and down of the water hours and hours after I leave the ocean."
"I think the ocean feels like a cool mistiness."
"I feel completely embraced, protected."

Not long ago I found myself opening a hatch on a white, smooth, and rather modern-looking capsule. If it hadn't been positioned in the corner of a rather small room, one might have thought it could take off like a quick electric car, ocean submersible, or some sort of flying craft. But it was none of those things. It was a very particular creation designed for a particular sort of quiet: a flotation tank. Inside the tank was warm, salty, 97-degree water—enough salt, 700 pounds, to keep my body afloat effortlessly at the water's surface. No bed is as comfortable as floating in warm water, but this was really something. The tank was big enough so that my body wouldn't come in contact with the sides or bottom; sound-insulated enough so that the only thing I'd hear would be the beating of my heart and the rise and fall of my breathing; and warm enough that the water, the air, and my body would feel like a continuum. And dark—as in no light whatsoever.

I took off my clothes, climbed in, closed the hatch, extended my legs and arms, and put my head back. I couldn't really tell the difference between having my eyes open or closed. I think they were mainly closed. Or open. Who knew? Not me.

The distinction between the water and the air was also unclear. For about thirty minutes, floating in the dark, my brain looped the familiar patterns, schedules, urgencies, absurdities. Then what I can best describe as a dissolving process began. Reference points, images, and ideas began to soften and erode like a tablet in water. My sense of location and time, my plans for the day, month, year, all slowly flaked away into nothingness. Eventually the sense of vast, timeless, open space in every direction took over.

Floating in warm water in quiet darkness was deeply relaxing. A quieter, larger womb? A primordial tropical night ocean? Hard to say. I know that I was in the isolation tank for ninety minutes because Shanti, owner of the Be Well Spa in the Santa Cruz Mountains, told me so when I emerged. But she could have said almost anything about the passage of time outside the tank and I would have believed her. ("Welcome to the year 2020, Dr. Nichols," she'd announce, handing me a glass of water. "Why, thank you, Shanti," I'd reply.) I came out moving slowly, taking my time as my senses reassembled themselves. My brain felt as if I'd been staring at the ocean for hours—the kind of "mindful mindlessness" that those deeply experienced in meditation spend years trying to achieve. I've been in and around water all my life, but never like this. I suddenly realized why "floaters" include everyone from software engineers, high-tech entrepreneurs, writers, actors, and other creative types, NFL players, and even (reportedly) U.S. Navy SEALs. That tank wasn't just a tube full of salty water—it was, well, what *was* it?[44]

Dr. John Lilly was one of the pioneers of research into the effects of isolation. "My research shows that when you eliminate external sources of danger, the inner experience... can be anything that you can allow yourself to experience," he wrote in *The Quiet Center,* "which can contain a great peace for those ready for it."[45]

Much like my own experiences, Lilly reports that for about the first forty-five minutes "the day's residues are predominant... gradually [the floaters] begin to relax and more or less enjoy the experience." But in the next stage the mind becomes both tense and bored. Researchers such as Dr. Donald Hebb at McGill University have looked at the effects of this kind of extreme isolation: subjects began to desire stimulation, were disoriented and confused, and eventually became delusional and hallucinated. (It's not surprising that Amnesty International classifies extreme sensory deprivation as torture.) However, if the need for stimulation is overcome through practice or brute mental effort, Lilly said, subjects in the tank pass into "reveries and fantasies of a highly personal and emotionally charged nature," followed by what he described as the opening of a "black curtain" into empty space. The theory is that flotation tanks allow the brain to transition from the waking state (beta waves) through the state of wakeful relaxation (alpha waves) and ultimately to the state of deep meditative consciousness akin to the moment between waking and sleeping (theta). In this altered state the mind settles into nothingness, the inner voice is silent, and often a feeling of oneness and bliss occurs—writer Seth Stevenson described it as "the closest you will ever come to having a drug-like experience without taking drugs."[46]

Flotation is being tested as a treatment for a wide range of physical, mental, and emotional ailments, including chronic pain, high blood pressure, motion disorders, tension headaches, insomnia, and so on. For some reason, much of the research on flotation is coming from Sweden, where it's been used to help highly stressed executives on the verge of burnout,[47] young people with ADHD, autism, PTSD, and depression,[48] stress-related physical pain,[49] and even whiplash.[50] Flotation has been shown to enhance creativity[51] as well as mental and physical performance.[52] You may never climb into a float tank. I may never find my way back to one. But the world abounds with simulacra, some quiet and others noisy, some big and some small, warm and cold, bright and dim. You will soon find yourself neck deep in water, and when you do, take some deep breaths, close your eyes, and go a little deeper.

5

Blue Mind at Work and Play

Out of water, I am nothing.

—Duke Kahanamoku (1890–1968), five-time Olympic swimming medalist, water polo player, and one of the fathers of modern surfing

Open-water swimming champion Bruckner Chase may not be as versatile as Duke Kahanamoku, but his exploits are still pretty impressive. Only the second person in history to successfully swim the 25 jellyfish-strewn miles across Monterey Bay in California, Chase also has swum the length and breadth of Lake Tahoe (22.5 and 12.0 miles), around Pennock Island in Alaska (8.2 miles) and three of the barrier islands in New Jersey (22.5 miles, 19.6 miles, and 16.5 miles), and across the channel from Lanai to Maui (9.6 miles). A retail operations executive by profession and endurance athlete since he was nineteen, Chase has turned his love of the water into a wide-ranging career: a trainer of lifeguards, a developer of open-water swimming programs for the Special Olympics, an ocean advocate sponsored by grants from the National Marine Sanctuary Foundation—and now, a coach to youth in American Samoa teaching teenagers how to become swimming, water safety, and ocean culture mentors. "Something profound happened to me when I started getting into the ocean," Chase said. "It keeps driving me to help others discover and embrace their own personal connection to water."

That personal connection is important to anyone who chooses a profession or sport that's in, on, under, or around the water. And that's a *lot* of people. In 2011 in the United States alone, for example, 21.5 million people did some kind of fitness swimming, 56.1 million Americans fished, 2.48 million people surfed, 2.8 million scuba dived, 9.3 million snorkeled, 3.8 million sailed, 7.5 million people jet-skied, 4.6 million water-skied, 10.17 million people canoed, 7.3 million kayaked, 1.38 million windsurfed, and 1.6 million went stand-up paddleboarding. In 2012 recreational motorboating added another *88 million*—a full 37.8 percent of the U.S. population.[1] Around the world, more than 500 million people choose water-based recreation as a means of exercise, escape, challenge, relaxation, excitement, and play. (The number is much higher if you include those who find spiritual connection via water. For example, tens of millions of people bathe in the Ganges River during the Indian festival called Kumbh Mela, the largest gathering of people for a religious purpose.)

On the professional side, the range of water-based work is equally impressive, from lifeguards to fishers to sailors to scientists, those who teach water skills, or supply seafood for our dinner plates, or provide and maintain the equipment and facilities needed for all forms of water-centric recreation, the men and women who serve their countries' economic and security interests in the navy, coast guard, and merchant marine. Some of the men and women in these professions face great danger, even death—as TV shows like *Deadliest Catch* and movies like *The Perfect Storm* and *Captain Phillips* remind us. People sailing the waterways can spend months away from dry land, in close quarters, working shifts that result in fragmented sleep and bone-draining fatigue. Why are some people drawn to work in an element that is far more lethal than earth or air?

And make no mistake about it: water is far more lethal. In 2012 only 362 people died in commercial airplane crashes worldwide. In the United States in 2012, there were a total of 1,539 crashes of any kind of civilian aircraft (commercial, commuter, private planes), with 447 fatalities. In contrast, the World Health Organization estimates that worldwide, there are 388,000 people who drown per year. This figure does not take into account people who drown in floods or as a result of marine transport.[2] The irresistible pull of water isn't true for everyone, of course; for some people their water-based professions are simply jobs, just as for some the laps they swim at the local pool are merely a convenient way to exercise. But read or listen to what white-water rafters and kayakers or swimmers or surfers or divers or sailors or fishers say about water, and you'll hear the language of people who are in love with, if not addicted to, their sport or profession.

In the Water: Swimming

This summer I swam in the ocean,
And I swam in a swimming pool,
Salt my wounds, chlorine my eyes,
I'm a self-destructive fool, a self-destructive fool.

— Loudon Wainwright III, "Swimming Song"

There are several reasons why swimming is the fourth most popular recreational activity in the United States, and the one people most aspire to take up if they're not swimming already.[3] First is convenience: all you need is yourself and a body of water. But it's the ways in which we interact with water that make swimming both healthful and enjoyable for body and brain.

It all comes down to viscosity, pressure, and buoyancy. Archimedes (the Greek philosopher who supposedly came up with the concept of water displacement while getting into a bathtub) stated that when an object enters the water, water moves out of its way. At the same time the water pushes upward against the object with a force equal to the weight of water displaced. That force creates buoyancy, or the ability to float. If an object is compact and dense (like a boat anchor, for example), it's heavier than the amount of water displaced, and it sinks. If an object is either light (like an inflated beach ball) or its weight is spread out over an area big enough to displace water equal to its weight (like the hull of a boat), it floats.

So why should the human body, which would seem pretty dense and compact for its weight, float? Easy—recall that the body (our blood, bones, organs, skin, and muscle) is up to 78 percent water when we are born (slowly decreasing with age), so we have close to the same density as the medium in which we swim. We're also around 15 percent fat, which is lighter than water, and we have lungs that are filled with air, which makes us more buoyant (like that beach ball). Therefore, a 200-pound human body actually weighs only around 10 pounds in water.[4] (This relative weightlessness is why, since the 1960s, astronauts have used water immersion to train for missions in outer space.) This also explains why such small flotation devices can keep us afloat.

But there's much more to the experience of swimming than buoyancy. Water has a tangible quality, a weight, and it has 600 times the resistance of air. Unlike earth or air, we can explore water in multiple dimensions—up, down, sideways; as neurologist Oliver Sacks comments, we feel tangibly supported and embraced by this "thick, transparent medium."[5] The resistance and pressure of water contribute to swimming's role as one of the best forms of both aerobic exercise and muscle toning. Because the pressure of water outside the body is greater than the pressure inside, explains Bruce E. Becker, director of the National Aquatics and Sports Medicine Institute at Washington State University, water forces blood away from the extremities and toward the heart and lungs.[6] The heart responds by upping its effort, pushing this extra volume of blood more efficiently with each heartbeat, and thus circulating upwards of 30 percent more blood volume than normal throughout the body. To cope with this increased load, the arterial blood vessels relax and create less resistance to blood flow.

Here's the intriguing part: one of the hormones that regulates arterial function is catecholamine, and catecholamines are part of the body's response to stress. As Becker describes it, "During immersion, the body sends out a signal to alter the balance of catecholamines in a manner that is similar to the balance found during relaxation or meditation." In other words, just being in the water can create a feeling of relaxation and a decrease in stress.

But that's not all. The lungs are receiving a greater volume of blood as well, which, combined with the pressure that water exerts on the chest wall, makes them work harder to breathe—approximately 60 percent harder than on land. This means that aquatic exercise can strengthen the respiratory muscles and improve their efficiency. In one study that compared aquatic aerobics with "dry" aerobics, Becker discovered that while various forms of aerobics improved fitness levels and some respiratory capacity, only aquatic exercise improved respiratory endurance. The muscles, too, are benefiting from the increased circulation as they receive greater amounts of blood and oxygen. And it's a good thing, too, because it requires effort to propel the body through water; in swimming, every muscle is benefiting from what is essentially resistance training (one of the best ways to increase both tone and strength). In addition, swimming works the large, smooth muscles of the body, stretching and lengthening the muscles, joints, and ligaments with each stroke, while the head and spine get a good workout with every breath you take. It all means that stroking through the water not only puts you into a psychologically relaxed state, but also makes you physically stronger.

Like other forms of aerobic exercise, swimming can produce the release of endorphins and endocannabinoids (the brain's natural cannabis-like substances), which reduce the brain's response to stress and anxiety.[7] Some theorize that the feel-good effects of swimming are related to the same "relaxation response" triggered by activities like hatha yoga. In swimming, the muscles are constantly stretching and relaxing in a rhythmic manner, and this movement is accompanied by deep, rhythmic breathing, all of which help to put swimmers into a quasi-meditative state. (We're going to talk a lot more about this state in a bit.) As one of the greatest competitive swimmers of our time, Michael Phelps, describes it, "I feel most at home in the water. I disappear. That's where I belong."

That sense of belonging increases with exposure. Recent studies have shown that regular exercise is associated with an increase in the number of new neurons in the hippocampus, the area of the brain linked to learning and memory.[8] More neurons means greater cognitive functionality; this may be the reason why regular aerobic exercise, like swimming, has been shown to help maintain our cognitive abilities as we age.[9] But there's something more going on with the exercise that swimming provides. Even though we spend our first nine months in "water," we are not born with the ability to swim. We talk about babies "learning" how to crawl, and then walk, and then run, but this happens without instruction. Our brains are designed for the natural emergence of these abilities. But the ways we use our bodies in water—having to time our breaths consciously, reaching up and over and pulling the water toward us, moving the legs independently of the pace that the arms are setting—is nothing like the way we move on land. We must *learn* how to swim, and this combination of cognitive effort and aerobic exercise has actually been proven to provide the greatest amount of what is called "cognitive reserve"—that is, the mind's resilience to damage to the brain.

Sadly, according to a 2010 study commissioned by the USA Swimming Foundation, 40 percent of Caucasian children, 60 percent of Hispanic children, and 70 percent of African American children have low or no swimming ability.[10] (Worldwide, drowning remains the leading cause of "unintentional injury" deaths among children under the age of five.) Given what we're learning about the physical and Blue Mind benefits of swimming, this isn't just a disappointment—it's a public health crisis.

And that's why open-water swimmer Bruckner Chase is in American Samoa, a small U.S. territory that sits 15 degrees above the equator and a six-hour plane flight from Hawaii. To help promote ocean awareness, in 2011 he swam the nine miles from Aunu'u Island to Pago Pago Harbor, something that no one had ever done before—not because it was a challenging open-water swim, but because all of the locals were sure he'd be eaten along the way. "This is a three-thousand-year-old culture, with a strong oral tradition," Chase remarked. "One shark attack two hundred years ago can be passed down generations and keep people from going into deep water. It's a culture that has ties to the sea, yet they had marine-patrol first responders who couldn't swim. Not surprising, the rate of drownings here is very high." Long memories (especially those involving sharp teeth) mean that fear of the water can transform a concern into a fundamental legacy. So Chase and his wife, Dr. Michelle Evans-Chase, set out to change the culture around swimming. They established a program called Toa o le Tai—Ocean Heroes—in which they trained teenagers to safely be in and around the ocean and then to teach younger kids the same skills. Chase is justly proud of the results of his Ocean Heroes, one of whom is a young man named Tank. "When I first talked to Tank, he wouldn't go in water over his head, period," Chase says. "Today, Tank is jumping into water that's one hundred feet deep and a year ago he thought was full of sharks. And his buddies are all begging to be part of the program."

My biological father, Jack Hoy, was a "water guy" and an avid, lifelong swimmer. Indeed, he was built to swim: wide shoulders, barrel chest, his body tapering cleanly to his feet. He competed on swim teams in high school and college, and throughout his life he swam, sailed, and fished. Jack's favorite spot in the world was Cape Rosier, and he returned to that part of Maine's rocky coast again and again. It was where his family and friends came together to hold his memorial service in August 2013. At Bakeman Beach, Jack's kids and grandkids stripped down at sunset and plunged into the cold Atlantic. We swam out into the cove, as he loved to do. I felt every stroke. It was cold, yes, but that didn't matter. Swimming together, in his name, here in this place, was perhaps the best tribute we could offer Jack. Somewhere out in the dimming deep water we naturally gathered in a circle. There really wasn't much to be said, the ocean and our exposed bodies—similar due to Jack's DNA but different due to diversity in his mates and spouses and those of his kids—spoke clearly for generations and ancestors, and we were all the same, connected by the thread, connected by water. A love of water was one of his gifts.

Some of the strongest recollections people can have around water are swimming alongside family. The dad who holds his arms out to you as you jump into the pool for the first time. The mom who sits with you on the beach in the shallow water, laughing with you as the waves rush around the two of you. The big brother or sister or cousin who leads you out into deeper water than you might ever attempt on your own. The love of swimming is often passed down through families, and being in the water together can bond you at every stage of life.

But perhaps water provided you with a different, not so happy, even phobic experience. (Two-hundred-year-old shark attack not required.) While my biological mother and father both had strong affinities for water throughout their lives, my adoptive mother has a lifelong fear of water, with a long list of illogical reasons for not getting wet, and would rarely get into a swimming pool with her kids (and never put her head underwater when she did). And she absolutely refused to go into any ocean, lake, or river. My adoptive dad was a little better around water, but no fish himself. Even competitive swimmer Leanne Shapton had a kind of fear of water. She writes:

> Being pool-trained, I'm used to seeing four sides and a bottom. When that clarity is removed I get nervous. I imagine things. Sharks, the slippery sides of large fish, shaggy pieces of sunken frigates, dark corroded iron, currents. I can swim along the shore, my usual stroke rolled and tipped by the waves, the ribbed sandy bottom wiggling beneath me, but eventually I get spooked by the open-ended horizon, the cloudy blue thought of that sheer drop—the continental shelf.

In—and On—the Water: Surfing

> *For a surfer, it's never-ending. There's always some wave you want to surf.*
>
> —Kelly Slater, champion surfer

Surfing has always been a part of the water studies at our annual Blue Mind conferences. I think it's because surfers probably exhibit more Blue Mind than anyone. They are attuned to the water, used to watching it carefully for hours on end, reading its changes, looking for the smallest indication that the next wave will be, if not the perfect wave, at least rideable. They are *in* the water as well as *on* it—they know the power of a wave to slam them down to the bottom, leaving them scraping the sand, rocks, or reef, holding their breath for dear life, fighting upward against the whirling energy to break through, gasping—yet still looking seaward for the next chance to hop on their boards and take the ultimate fifteen- or thirty- or sixty-second ride.

Recently I've been surfing with my buddy Van Curaza near Santa Cruz. Van is addicted to black coffee, helping people, and waves—and he's among the best in the world at all three. Van has ridden those waves since he was ten years old. He surfed huge breakers off the coast of central California; for a while he turned pro and was sponsored by top brands like O'Neill and Quicksilver. He admits freely that he was drawn to the "aggressive, adrenaline-filled atmosphere" of pro surfing and its lifestyle of too-frequently-associated habits: drinking, pot, coke, prescription pills, meth. Van's been sober since December 2002, but back when he was an elite surfer, he didn't realize that the very parts of his brain that were drawn to (and being shaped by) surfing were also producing that addictive urge.

Howard Fields, whom we met earlier, and Dr. David Zald study the neurochemical mechanisms of addiction and substance abuse, and they both see clearly how these mechanisms are at play in surfing. "Addiction occurs when dopamine neurons and the nucleus accumbens (the "pleasure" center of the brain) are stimulated by certain actions . . . and the brain computes the value of such an action to optimize future selection," Fields has written. Remember that dopamine release is associated with novelty, risk, desire, and effortful activity; it's also a key part of the system by which the brain learns. All of these factors, Zald points out, are present in surfing: "As surfers are first learning, there's an amazing burst of dopamine simply when they stand on the board—'I didn't think I could do that!' And then surfing is never going to be exactly the same. The wave comes, but it's always somewhat unpredictable." Novelty? Check. Risk? Check. Learning? Check. Aerobic activity? Check. Dopamine? In spades.

But that's not all. As we discussed above, aerobic exercises (such as surfing) produce endorphins, the opioids that affect the prefrontal and limbic areas of the brain involved in emotional processing, and create the feeling of euphoria known as runner's high.[11] The beauty of the natural environment where people surf also increases the sense of a peak emotional experience. Add the dopamine, the endorphins, and the natural setting to the adrenaline rush produced by the amygdala's "fight or flight" impulse when a surfer is faced with a large wave (or a wave of any kind when you're first starting out), and you've got a seriously addictive experience.

Unfortunately, the brain starts to crave this dopamine–endorphin–adrenaline cocktail, and it looks for other ways to produce that same feeling. As one surfer described it, "Once you surf, it's about the next wave you're going to catch, because you always want more, like a drug." That's why many surfers like my pal Van get drawn to other risky pastimes, including drugs like cocaine, meth, alcohol, and nicotine, all of which flood the dopaminergic system with an instant hit of pleasure.[12]

But what if this need for dopamine stimulation could be turned to good? What if it's possible for people to substitute surfing for other, more destructive addictions? That's exactly what Van has done; one of his motivators to quit drugs and alcohol was the thought, *If I keep doing this, I won't be able to surf anymore.* Since 2008 Van has run a program called Operation Surf, in which he shares his love of waves with anyone who stands still long enough for him to get them into a wet suit, onto a surfboard, and out on the ocean. His clients are so-called at-risk youth, people with terminal illnesses or physical limitations that you might imagine would preclude surfing, veterans suffering from PTSD, and folks with surfing on their bucket list. He helps addicts like himself find a different kind of high composed of wind, wave, and ocean—and camaraderie like you've rarely witnessed.

It's extraordinary to see Van's clients after a session on a board. Many of them are noticeably calmer, quieter, and happier. The theory is this: once the adrenaline has worn off and the dopamine storms in the brain have calmed down, the body is still producing endorphins (runner's high usually doesn't kick in until after thirty minutes of aerobic exercise, and can last up to two hours afterward). It's also been saturated with an abundance of negative ions, which have been shown to lower blood lactate levels and elevate mood. Surfers often report feeling calmer and happier after a session on the water.[13] Van's greenhorns certainly look as if they're experiencing the Zen-like experience that's called surfer's stoke.

Back on the beach after that morning session in Santa Cruz with Van and Operation Surf, one of the instructors told me that he really didn't like surfing on days like this, when the ocean is so cold (49 degrees), but "this feeling, right now, makes it worth it." Out of the water, there's nothing like a hot cup of black coffee and some surf talk about the day's best waves.

With these kinds of emotion-packed experiences, it's no wonder that surfing generated more than $6.24 billion in revenue in 2010 and is one of the most popular water sports in the world. It's practiced on the coasts of six of the seven continents, by people of every race, color, and creed; it's estimated that 23 million people surf worldwide, more than 1.7 million in the United States alone.[14] Even in some of the most challenging circumstances, young men and women are taking up boards and figuring out how to catch a wave, and Farhana Huq is an advocate and champion of young female surfers worldwide. Farhana is a powerhouse: medium height, a warm smile and ready laugh, and a background in dance and martial arts—she was the first South Asian girl on the USA Karate national team. She is used to breaking barriers: her parents are successful Pakistani/Bangladeshi immigrants whose traditional cultures do not usually encourage women to participate in such sports as karate and surfing. But Farhana grew up around surfers in New Jersey, and at age twenty-six, while on a trip to Hawaii, she decided she wanted to learn to surf. "It was the hardest sport I ever tried," she recalls, but it didn't take long for her to become as addicted to surfing as had Van Curaza. When she wasn't running C.E.O. Women (the nonprofit she founded to help low-income immigrant women become entrepreneurs), she started traveling to surfing spots around the world. Instead of simply looking for the next wave, however, Farhana looked around these far-off locations. "I realized that women and girls in the most unlikely places—in Bangladesh, in India and even along the war zone of the Gaza Strip, all very conservative cultures—were starting to surf and were the first to pursue an ocean-loving sport. I wondered about these trailblazers and what gave them the strength to take these risks," she said.

When she came back home she founded Brown Girl Surf, to honor trailblazing females who are using surfing to empower other women and girls and their communities to be stoked about just being in the water. For example, Nasima Atker from Bangladesh was homeless when she started surfing. Since then she has braved the discouragement of male surfers on the water as well as her community and peers on land to train to become her country's first female lifeguard. With Brown Girl Surf, Farhana wants to connect with and encourage women around the world to discover the sport that has given her so much. "I had a health scare a few years ago, and my first thought was, *If this is serious I want to get in as much surfing as I can now in case I don't make it,*" she said. "It made me realize how much I equated surfing with living."

Farhana is quick to point out that "surfing isn't the solution to all one's woes or social problems . . . it's not a cure per se, but there is definitely something there . . . something that it gives us." In her case, that "something" is very positive indeed.

Under the Water: Diving

> *From birth, man carries the weight of gravity on his shoulders. He is bolted to earth. But man has only to sink beneath the surface and he is free.*
>
> —Jacques-Yves Cousteau

Jacques Cousteau's grandson, Fabien, has been heading deep underwater since he was four years old. He's scuba dived with humpback whales and sharks, and done free dives with sea lions. This year he plans to spend an entire month at the bottom of the ocean off the Florida Keys in the aqualab Aquarius Reef Base. But Fabien is all too aware of the dangers: his cousin Philippe Cousteau was diving with naturalist and TV star Steve Irwin when a stingray fatally stabbed Irwin in the heart, and his family has experienced countless near-miss ocean accidents through the decades.[15] Nevertheless, Fabien, along with dive enthusiasts around the world, is drawn again and again to plunge beneath the surface of seas, oceans, and lakes and experience what the American painter Robert Wyland (known simply as Wyland) called "the world's finest wilderness."

If swimming provides a novel environment for the human body, diving is truly entering another world, one that teases yet warps our senses. Submergence puts pressures upon our cells that they were never designed to encounter. Sight and hearing function differently underwater; your movements are slowed, delayed, elongated, magnified; you are literally pushing the water out of your way with your hands, feet, and body as you glide along; colors are distorted, distance perception is uncertain, sound (what there is) travels faster and can seem to come from everywhere at the same time.[16] Most of all, we move underwater in a completely unaccustomed way: we are able to hover horizontally above a fish or plant simply by using our breath and a little movement of our fins to keep us stationary. One remarkable woman in Great Britain, Sue Austin, has even turned the unique properties of underwater movement into an art project: she performs underwater ballet while diving in her wheelchair.[17]

As geographer Elizabeth R. Straughan points out in her study of diving, touch, and emotion,[18] it is the tactile sense of being in and moving through water that provides some of the most interesting physical aspects of diving. "Divers are always touching," she writes. "They are supported, suspended, moved and compressed by the water that encompasses them. . . . It is through somatic [of and relating to the body, as distinct from the mind] tensions and pressures felt within and around the body that divers experience a material and tangible connection with the water environment."[19]

Psychologist David Conradson once defined stillness as "an internal state of calm in which a person becomes more aware of their immediate embodied experience of the world and less concerned with events occurring 'out there.' "[20] So it is on a dive: the surface world, with all its obligations and concerns, seems far away, even somewhat unreal. "I'm only forty feet below the everyday world, but I might as well be light years away," one diver said. The usual noises of the outside world disappear, to be replaced by the steady sound of your own breath in the regulator, the occasional muted grunts from a fish, snapping of shrimp, or bubbles of another diver nearby."

Yet any kind of diving carries with it risk—not just of drowning, but also of having to adjust to the (literal) pressure placed upon the body. The deeper you descend, the greater the pressure—one atmosphere's worth for every ten meters—and the more the water compresses the gases inside your body, causing your tissues to absorb more nitrogen, which then has to dissolve out of your tissues as you ascend. Go too far down and the nitrogen enters your cardiovascular and central nervous system, making you light-headed and maybe a little dazed from "nitrogen narcosis." Ascend too quickly and the nitrogen bubbles don't have enough time to leave your body, causing the bends. There are also dangers linked to holding your breath underwater, having panic attacks (experienced by almost half of all divers, according to some studies),[21] problems with the inner ear, and even issues with too much CO_2 in the bloodstream.[22] Recent studies also indicate that breath-hold—"free"—diving in extreme conditions may be a contributing factor in the formation of white-matter lesions in the brain.[23]

Still as with surfing, the awareness of potential danger as well as the opportunity to challenge oneself in an unfamiliar and ever-changing environment can be the very things that attract people to the sport. Divers tend to have a greater appetite for adventure, to be somewhat more aggressive, less inhibited by and prone to anxiety, and healthier.[24] And it's very likely that diving, like surfing, is a way for stimulation-craving individuals to get their emotional/dopamine "fix" and come back to the surface calmer and happier.[25] Under the water, divers can push their limits; explore places that are always new, exciting, different, and slightly dangerous; and share these peak experiences with others who are just as enamored of the sport as they are. And if they can do all of this in the middle of the world's largest, most beautiful, yet unexplored environment, so much the better.

Those surfers hoping to manage waves the size of four-story buildings and the divers hoping to snatch pearls from depths even greater are but a tiny minority of the millions who take to the sea each year. Fortunately for those not interested in risking the bends or a snapped surfboard, it's still possible to reap the astonishing benefits of immersion. You don't have to risk your life to improve your life—though sometimes a fish gives up its own along the way.

By the Water: Fishers

Many men go fishing all of their lives without knowing that it is not fish they are after.

— Henry David Thoreau

I've spent a lot of time walking coastlines, lakeshores, and riverbanks in the United States and Mexico. Everywhere I travel, I always seek out the water and spend time strolling along its edge, observing it and speaking with the people who love it. I've often seen fishers there—men, women, and children—poles in hand or planted in the sand or grass next to them. They're scanning the water intently, searching for the splash or stir or wavelet that means a fish may be close to the hook. Or perhaps they're chatting with the fisher next to them, not really paying attention to the task at hand. Maybe they're simply looking out over the water and enjoying the beauty of the setting. Or perhaps they've got their eyes closed, taking a snooze in the middle of a lazy afternoon. I guess it's not surprising that some of the earliest books on fishing, dating from the fifteenth century onward, talk as much about the personal rewards of fishing as they do about angling techniques.[26]

Frequently, among the heartening things I see are children with dads, moms, grandpas and grandmas sitting or standing next to them, teaching them how to fish. As a dad, one of my greatest pleasures and responsibilities is to pass along the love of all things water to my daughters. And it's pretty clear from both anecdotal and sociological research that being introduced to fishing in childhood is a prime indicator that a person will be fishing in adulthood.[27] It doesn't matter what kind of fishing you enjoy—ice fishing, fly-fishing, deep-sea fishing—as Dan Pearce wrote in *Single Dad Laughing,* "Fishing is much less about the fishing, and much more about the time alone with your kid, away from the hustle and bustle of the everyday."

Because fishing is so popular,[28] there have been several studies in both the United States and Australia investigating the motivations of anglers. Turns out that Thoreau got it right: catching a fish was *way* down the list of reasons for going fishing. In Queensland and the Great Barrier Reef region of Australia, for example, fishers mentioned the desire for rest and relaxation, to get away from everyday life, and to be outdoors, in a natural environment, as the most important factors of a satisfying fishing trip.[29] In Minnesota, a quality environment and a sense of freedom were keys to a great fishing experience, not what ended up on the hook.[30] Of course, some of the people I've passed on my walks wanted (or needed) to catch their dinner. But I imagine that even for them, the chance to do so next to water, in nature, and to focus simply on one thing and one thing only, gave them a different kind of nourishment.

Think about what's required to fish. Not the physical equipment—that can be as basic as a pole, string, a hook, and some bait—but the physical and mental wherewithal needed: focus and clear thinking. A steady hand and the ability to handle emotions like elation, disappointment, and boredom. Patience—boy, do you need patience—yet also the ability to respond at a moment's notice to a tug on the line when a fish snags the bait. And, as in scuba diving, you also need the ability to slow down and appreciate the beauty of the watery setting in which you find yourself. If you do, then you can experience some of the health benefits reported in an Australian study that showed the process of preparing for and then going to fish (preferably with other fishers) can quiet the mind, improve cardiovascular health by reducing stress, and create a general sense of well-being.[31]

Not surprisingly, fishing of all sorts is used quite successfully as recreational therapy. For example, the casting motions used in fly-fishing are ideally suited to help women recovering from breast cancer surgery to rebuild, strengthen, and increase flexibility of arm and chest muscles, since lymph nodes, muscle lining, and even small chest muscles are often removed in the procedure. Recreational fishing's pace and calm means it can be enjoyed by numerous people with physical and mental disabilities, as well as those suffering from traumas like PTSD or traumatic brain injury (TBI). One program, called Fishing 4 Therapy, in Sydney, Australia, gives individuals with cerebral palsy, Down syndrome, visual impairment, brain injury, hyperactivity, or motor skills difficulties, and their families, the opportunity to enjoy an afternoon's fishing trip. Sure, it takes some adaptation of equipment and a little patience, but for many of the program's clients, fishing is one of the few ways for them to get outside and to participate in recreational sport. The barriers to entry are lower and the basic skills are rather simple. For some clients, "just the act of holding a fishing rod and attempting to turn the reel handle can be a major therapy accomplishment," according to the authors of a 2011 Australian report on the benefits of fishing. "Specific benefits include: a positive impact on mental health; perceived well-being; prevention of chronic disease; and reductions in health care burden."[32] Being outside, in motion, with those we love, engaged in a new activity while floating on or standing by water sounds idyllic no matter where or who you are.

Ultimately, fishing is a water sport that can be enjoyed by anyone who is old enough, or not too old, to hold a pole. "It is a very democratic kind of activity," pop-culture pundit Faith Salie wrote. "You don't need to be tall or strong or agile. You just need to be patient, or drunk... on the beauty of nature. It's kind of a metaphor for a good life. Try your best, hope for the best, have days when you catch something and days when you don't, but always, always be thankful for the sound of the water and the sun in the sky and the chance to cast another reel."[33]

On the Water: Boaters, Paddleboarders, Kayakers, and Canoers

> *Believe me, my young friend, there is nothing—absolutely nothing—half so much worth doing as simply messing about in boats.*
>
> —Kenneth Grahame, *The Wind in the Willows*

According to the Outdoor Industry Association, 1.24 million people tried stand-up paddleboarding last year, up 18 percent from 2010. Sales of stand-up paddleboards doubled between 2010 and 2011. While stand-up paddleboards have led to a fitness craze only in the past decade, in truth they're the most ancient and basic form of boat. Our distant ancestors probably used boats that were not much more sophisticated to journey from island to island in Southeast Asia and then on to Australia more than 50,000 years ago.[34] Of course, nowadays most boating—from self-propelled personal watercraft such as paddleboards, rowboats, rafts, kayaks, and canoes, to wind-powered sailboats of every size, to engine-powered jet skis, motorboats, and yachts that require multiperson crews—is less about transportation and more about enjoying the water. If you talk to recreational boaters about why they chose their particular sport, you hear some of the same reasons that fishers and surfers offer: it's the chance to be on the water in a very intimate way. Those who use self-propelled vessels also extol the health benefits, the cardiovascular exercise, the upper-body workouts of paddling for hours, the decidedly meditative state that arises with the rhythmic strokes of paddle dipping into water. But motorized or not, a boat offers a chance to get away from it all, to immerse yourself in the sights and sounds and the feel of whatever body of water you're on, to enjoy the fresh air.

With larger sailboats and other vessels that require more than one person, you hear about teamwork, trust, and shared adventure. Research shows that the combination of gaining self-confidence while working in a team is valuable for troubled youth. Today programs around the world use boats and sailing as rehabilitative therapy for those with physical disabilities (including paralysis, blindness, deafness, and amputation); developmental disabilities like ADHD, autism, and Down syndrome; those with traumatic brain injury (TBI) and other injuries; as well as people who have experienced emotional trauma. In Newport, Rhode Island, and Nantucket, Massachusetts, Sail to Prevail has a fleet of specially adapted sailboats, ranging from twenty to sixty-six feet, in which more than fifteen hundred people with disabilities learn the basics of sailing. Vessels include *Easterner*, a one-of-a-kind sixty-six-foot handicapped-accessible America's Cup racing boat. Sail to Prevail reports significant improvements for its clients: 91 percent have more confidence, 90 percent feel they increased their teamwork skills, and an incredible 99 percent say they have a more positive outlook on life. No doubt the super-neurohormone oxytocin is at play during these novel, enjoyable, yet mildly stressful experiences, fine-tuning the brain's social instincts, priming participants to crave social contact, enhancing empathy, and increasing willingness to be helpful and supportive. Of course, there are many other ways to develop teamwork, but the water factor adds remarkable potency to the effort. So too does a quality you hear mentioned most often from boaters of all kinds when they describe their reason for taking to the water: *freedom*. In a boat, you feel as if you are "the master of your fate, captain of your soul," to paraphrase William Ernest Henley. For children who through no fault of their own are forced to live a difficult life, constricted by circumstance, freedom is a profoundly uplifting liberation.

Just like marathon open-water swimmers and big-wave surfers, there are those who choose to take that sense of freedom and independence to the extreme—by rowing or sailing solo across the oceans and sometimes around the world. There are two premier circumnavigation races for sailing yachts: the Velux 5 Oceans race (sailed in stages) held every four years, and the Vendée Globe, sailed nonstop around the world. These races put incredible strains on the men and women who sail; the nonstop stress of storms, high seas, difficult conditions, food and sleep snatched whenever possible (and never enough of either). And above all, the isolation—anywhere from 78 to 1,152 days spent at sea, alone. Extended isolation is both psychologically and emotionally taxing, even destructive, producing hallucinations, depression, anxiety, impaired thinking, rage, and despair. In a 2012 experiment, the brains of rats isolated for eight weeks had a reduced ability to produce myelin in the area responsible for complex emotional and cognitive behavior.[35] (Myelin coats and insulates axons in the nervous system.) A 1968 solo nonstop circumnavigation race had nine competitors, only one of whom finished the race. Of the remainder, five dropped out after just two months at sea, one killed himself after he returned home, another was believed to have committed suicide before he landed (he had spent most of the contest holed up by the coast of Brazil rather than racing, and was terrified his deception would be found out), and the final competitor abandoned the race and turned up several months later in Tahiti.[36] But despite all the dangers and obstacles, something draws these men and women to boats and the sea: "... sailing is a dance, and your partner is the sea," writes Michael Morpurgo, author of *Alone on a Wide Wide Sea*. "And with the sea you never take liberties. You ask her, you don't tell her. You have to remember always that she's the leader, not you. You and your boat are dancing to her tune."[37]

That music is sweet in its melody. And for every sailor destroyed by extreme conditions, there are hundreds of millions who discover that the ocean not only resuscitates but also enhances.

Working the Water: Professional Watermen and -women

> *There comes a time in a man's life when he hears the call of the sea. If the man has a brain in his head, he will hang up the phone immediately.*
>
> — Dave Barry

It takes a special kind of person to make a living from being on the water—maybe just a little bit crazy, or a little bit obsessed, or a lot in love with the big blue. I've certainly seen all three among the professional watermen and -women I've hung around with over the years. I'm not talking here about those fortunate few who have turned their love into a professional sports career (rare indeed) or become a teacher, coach, or supplier of equipment for water sports. Nor am I talking about my fellow scientists who study plants, animals, weather, psychology, physiology, oceanography, and a wide range of other -ologies related to water. And I'm leaving out those who sell or prepare seafood or aquatic plants for humans to eat. The ones I mean are the sailors who steer barges through canals and rivers, or who work the huge ships that constantly cross the lakes and oceans of the world. I'm also talking about the commercial fishers (including those who trawl for or trap shrimp, mussels, clams, and crabs) who go out on boats rather than run fish farms to feed their families by feeding us. These men (and a few women) have chosen careers that take them far away from homes and families, sometimes for weeks and months at a time, in difficult conditions that put them in danger of injury or death. (While maritime disasters are relatively rare and in decline, each year approximately one hundred commercial ships sink. And commercial fishing still ranks as the most dangerous profession in the world.)[38]

You see many of the same stressors that solo sailors face in the lives of those who make their living from water. They are subject to all the vicissitudes of weather, wind, and wave. Boats are designed to stay upright, sure—but the amount of energy produced by some waves is sometimes more than even the biggest ships can handle. Water also transfers energy more than land or air, so the swells of the sea combined with gusts of wind make any storm very, very serious. It also gets cold on the water, and the dampness of spray makes you feel colder still. The combination of chill, wet, waves, corrosion, and the kind of heavy equipment common on ships and fishing boats is incredibly dangerous. (Jeff Denholm, a "Surf Ambassador" for the clothing company Patagonia, lost most of his right arm on a fishing boat in Alaska—the boat lurched, and his arm was sucked into an exposed driveshaft. It took twelve hours to get him to the nearest medical center.) Plus, as any sailor or fisher will tell you, on most voyages sleep is in short supply and bone-wearying fatigue is all too common. Because many of a ship's critical jobs (for example, navigation and engineering) must be manned 24/7, most large ships divide the day into watches of anywhere from three to five hours, and sailors will stand at least two watches every day, usually in the pattern of three hours on, nine off, or four hours on, eight off, and so on. This means that most mariners get to sleep less than five hours at a time, and they may have to stand watch in the middle of the night one day and in the middle of the afternoon the next. Chronic lack of sleep adversely affects the brain and cognitive function. Researchers at the Stanford Center for Sleep Sciences and Medicine advise that the average amount of sleep needed to avoid the adverse effects of sleep deprivation is about eight hours. Of course, being a fisher can be even worse. Many fishing boats stay out until they have gotten their catch, whether that takes a few hours or much, much longer. Then they turn around and go out again. As one fisherman remarked, "You goes three to four o'clock in the morning, and you don't be in till one or two in the day. When you gets in then you gets rid of your fish . . . you're baiting up till seven in the evening."[39]

So what makes sailors and fishers stay on the water? Like many tradition-based professions, jobs are often passed down from generation to generation. In a 2000 study of fishers in British Columbia, eleven out of twelve had close relatives who were fishers. In some places—the small coastal villages in Baja California, Mexico, where I studied sea turtles, for example—fishing or maritime jobs are the main sources of income and, often, food. In many poorer countries, a job on a ship is seen as an opportunity to earn more money than could ever be possible at home. But ultimately, when you sit with fishers and sailors and talk to them about their lives on the water, after the tall tales and the grousing about shrinking fish supplies or stingy owners and corrupt middlemen, or bad food or little sleep, you hear something else: a pride in their work, an enjoyment of the adventure of entering a challenging environment every day, and a pleasure in the sense of freedom and occasional awe that being on the water gives them. Many of them, frankly, can't imagine doing anything else. Even many fishers in Japan, whose homes and boats were destroyed in the 2011 tsunami, returned to the water because, as one named Koichi Nagasako said, "I love being on the boat. I love the sea and I love fishing."

But there's a problem. Overfishing and inefficient fishing in many rivers, lakes, and seas have decimated stocks and forced many governments to put restrictions on catches—or to ban fishing altogether in some locations. As a result, the number of maritime jobs available is steadily decreasing. In 2013 the New England Fishery Management Council, which oversees fishing in an area once famous for the abundance of its cod, had to cut back catch quotas by 61 to 80 percent in an effort to restore decimated cod stocks. In many locations, fishing boats that could make a good living now struggle to catch enough to keep themselves in business. Luckily, fishers are coming together to save their jobs by changing their ways. Some have shifted from trawling (which produces a lot of collateral by-catch, as they call species of fish and other creatures that are not their targets) to using hook and line or traps, which net a smaller number of fish but usually more commercial varieties. Globally, in places like Mexico, India, and China, by-catch remains an enormous problem with known but politically untenable solutions. But in Morro Bay, California, a group of individual fishing boat owners have joined with government agencies and the Nature Conservancy in "a collaborative model for sustainable fishing." Boats are given quotas of how much of different fish stocks they are allowed to catch. If they catch too much, they report where the catch occurred so that other boats will know to stay away from that area. The partnership helps to protect overfished species, gives fishers known targets of what they can catch, and helps to keep by-catch to a minimum.[40] Fishers are saving their livelihoods while protecting fish stocks for the future.

Years ago, I saw a version of this same process along the bays and beaches of Baja California Sur. I was a graduate student who'd been told by his advisors not to bother trying to study sea turtles in Mexico because their numbers had already been decimated by years of poaching, pollution, overfishing, and habitat destruction. Most experts considered it too late to save the black sea turtles of Pacific Mexico. But I was young and idealistic, so my friend Jeff Seminoff (now leader of the Marine Turtle Ecology and Assessment Program at the National Oceanic and Atmospheric Administration, the federal agency focused on—surprise—the oceans and atmosphere) and I hopped into my sun-faded 1972 International Harvester Travelall and hit the road. We drove from Tucson down the long spine of the Baja California peninsula and the Pacific coast of mainland Mexico to visit the waters where turtles lived and grew, and the beaches where sea turtles had come to lay their eggs for millennia.

This was not the tourist part of Mexico's Pacific coast. Bahia de los Angeles, Punta Abreojos, Santa Rosalita, and Bahia Magdalena are areas occupied by small fishing communities inhabited and connected by people whose fathers, grandfathers, and great-grandfathers made their livings from the sea. Over the course of two decades we talked with the local fishermen, made friends of them, hired them to teach and ferry us—and later our students—out into the ocean in their *pangas* (twenty-two-foot fishing boats with little outboard motors) to find the turtles as they basked on the surface or foraged in the sea-grass beds and tag them so we could track their movements. We explained that we were not with the government, not there to catch turtle poachers or to make trouble for anyone. We simply wanted to measure and track the turtles and learn more about them.

But in truth, I also had another motive: I wanted to change the way people thought about turtles as part of an effort to ensure their continued existence (both turtles and fishers). It seemed like a long shot, and all of the turtle experts I consulted in the States and Mexico considered it a lost cause. In these small coastal communities, sea turtles were considered an important food source and part of the traditional local diet. The fishermen who didn't eat the turtles they caught could make a tidy profit selling turtle meat illegally to consumers and restaurants in other parts of Mexico. Anyone who might even mention the idea of protecting turtles instead of eating them faced accusations of eliminating a mainstay of the local way of life. And as is the case for many fishermen around the world, the idea of abandoning the only way they knew how to make a living—as well as a profession they loved—was not an attractive, or even a feasible, option. Early on I realized that if I'd been born in their town, I'd be doing the work they did, and if they'd been born where I was, some would likely have become marine biologists. (We certainly overlapped in our devotion to watching *Animal Planet* on TV.)

We started small and slowly, because we knew we needed to understand the fishers' world and its practical constraints. These men were not our opponents, nor were they malicious; they were people just like us, and we all needed to be able to look into each other's eyes and see our common humanity. Juan de la Cruz Villalejos, Julio Solis, Rodrigo Rangel, Francisco Mayoral, Isidro Arce, Javier Villavicencio, and Jesus Lucero—we fished with them and stayed in their homes. We had long conversations, asking them about their lives and occupations. We learned about the challenges of supporting families on what these men could catch, and yet we also discovered their enormous love of the sea and pride in their profession. (At one point I even ate turtle and attended a traditional turtle slaughter with them—pretty tough for a turtle biologist.) We agreed about the beauty of this part of Mexico, with its beaches and abundance of wildlife.

As we worked with them, talked with them, lived with them, we developed a sense of mutual respect and trust with their community. With that trust, it was possible to start asking the question "How can we work together to protect the turtles so they will be there for your sons and daughters, and their sons and daughters?" After several years of steady collaboration, we asked dozens of the fishers from the villages around the region to come together and share their extensive knowledge of turtles and their habitats. We talked about the reality of a dwindling number of turtles and the fact that if something wasn't done, there would be no more turtles to catch. With the backdrop of an unreliable, corrupt government and severely limited funds for environmental enforcement, a creative and participatory approach was the only solution. What if they agreed not to stop eating turtles, but to cut back on the number they caught each year? Could they help us protect the turtle population and habitat? Maybe there were other ways to earn a living from the sea that didn't require fishing many species to extinction or causing enormous yet unintentional harm by using nets that catch many nontarget animals.

It has taken more than twenty years, but those simple questions and, more important, the relationships built with Mexican fishermen and *costeños* (people of the coast) have resulted in the protection of tens of thousands of adult sea turtles in the ocean, restoration of miles of turtle-nesting beaches, an enormous reduction in sea turtle deaths, and strong signs of a revival of the numbers of sea turtles, from the sea-grass beds of Baja to the nesting beaches of Michoacán. Many of the fishermen have become dedicated conservationists and are helping preserve the natural wonders *and* traditions of their communities. (One of them, Julio Solis, is executive director of a local chapter of Waterkeepers that is fighting unsustainable development projects in Baja California while championing the causes of clean water and restoring sea turtles.) Others are participating in a thriving ecotourism industry, leading trips that allow people from around the world to get "up close and personal" with turtles and other marine wildlife. Still others, such as Grupo Tortuguero and the indigenous Nahuat community of Colola, Michoacán, are inspiring the next generation in their villages by speaking in schools, holding sea turtle festivals, and taking children on overnight turtle monitoring expeditions. I am grateful that they can link their new professions with the love of water they've had all along—the love that's shared by those who enjoy water sports (and water jobs) of all kinds.

Water sports and water careers build better brain chemistry and confer therapeutic benefits on both well-bodied and disabled (emotionally, genetically, experientially, and environmentally) alike. For those of us fortunate to live, recreate, or make our living on or near water, this wellspring of health, well-being, and therapeutic utility may be obvious, mildly apparent, or waiting to be illuminated. What is clear is that the more we recognize and appreciate what we have, the more we can multiply the benefits we can derive. This is a new kind of water cycle, one that tracks water's cognitive and emotional services, generates new conversations, businesses, and policies, and, together, an even greater affinity for the water. For my colleagues in Baja, the virtuous circle is playing out in practical terms. A shared love of the ocean built a successful and diverse conservation team. The survival of many more sea turtles enhanced a sense of pride and teamwork. The return of the animals promoted ecotourism and jobs. Young Mexican biologists and guides helped to grow the ranks of the pro–sea turtle movement even further. As my dear friend Chuy Lucero, a fisherman turned leading turtle-saver, is fond of saying, *"En el mar la vida es más sabrosa"*—life by the sea is sweeter. We have pledged to each other that one day when we are both great-grandfathers, we'll gather our clans on that shore and raise a toast to the sea turtles.

Adrift

Today, when we talk of conservation, the assumption is that we're talking about trying to save sea turtles, old buildings, endangered ecosystems, electricity, and so on. But we don't think enough about a more internal conservation, conserving our attention for what matters, conserving our engagement for what's important, conserving our acuity for decisions that make a positive difference. If you don't know what you want to save, you can't know how to save it. In the next section of this book we're going to look at what we're losing, and how to bring it back—and not surprisingly, that restoration is all about getting your Blue Mind on.

6

Red Mind, Gray Mind, Blue Mind: The Health Benefits of Water

Water symbolizes many things connected with healing. Pouring forth from within the depths of the earth, it represents life and regeneration.

— Wilbert M. Gesler, *Healing Places*

One of the most famous first lines of an American novel consists of three simple words: "Call me Ishmael." But in the rest of the first paragraph of *Moby-Dick,* Herman Melville's protagonist describes the power of water to heal his mental state:

> Some years ago—never mind how long precisely—having little or no money in my purse, and nothing particular to interest me on shore, I thought I would sail about a little and see the watery part of the world. It is a way I have of driving off the spleen and regulating the circulation. Whenever I find myself growing grim about the mouth; whenever it is a damp, drizzly November in my soul; whenever I find myself involuntarily pausing before coffin warehouses,

> and bringing up the rear of every funeral I meet; and especially whenever my hypos get such an upper hand of me, that it requires a strong moral principle to prevent me from deliberately stepping into the street, and methodically knocking people's hats off—then, I account it high time to get to sea as soon as I can. This is my substitute for pistol and ball.

Ishmael seeks the solace of water to bring him out of the "November" of his soul. Melville's sailor is a fictional character, but what he was onto is very real, indeed.

Yet instead of going to sea to manage our stress, like Ishmael, today we turn to the easier (but less effective) path: pharmaceuticals. As of 2013, according to the Mayo Clinic, antidepressants were the second most commonly prescribed class of drug in America, and opioid painkillers (which can be exceedingly addictive) were third.[1]

Making things even worse, analysis of FDA clinical trials show that at least four of the main selective SSRI—serotonin reuptake inhibitor antidepressants—don't perform significantly better than placebos in treating mild or moderate depression.[2] We're taking (and paying for) millions and millions of drugs, both prescribed and self-administered, that do little more than offer the promise of addiction—one *more* thing to be majorly stressed about.

Monkey mind. Toxic stress. Chronic stress. Stress overload. Directed attention fatigue. Mental fatigue. When sustained over long periods of time the "always on" lifestyle can (and will) eventually result in memory problems, poor judgment, anxiety attacks, nervous habits, depression, loss of sex drive, autoimmune diseases, and overreliance on alcohol and drugs for relaxation in all but the most genetically gifted among us. Chronic stress damages the cardiovascular, immune, digestive, nervous, and musculoskeletal systems. It lowers levels of dopamine and serotonin, causing us to feel exhausted and depressed. Studies have shown that stress that lasts longer than twenty-one days can impair the function of the medial prefrontal cortex[3] (which affects higher-level thinking), while it makes the amygdala (the fear and aggression center of the brain) hyperactive.[4] Prolonged exposure to gluocorticoids, a type of steroid secreted by the adrenal glands during stress, can cause the cells in the hippocampus to atrophy—the same damage seen in people with post-traumatic stress.[5] A 2012 report concludes bluntly that consequences of prolonged stress include an increased risk of premature mortality.[6] Stress really does kill. And yet, off we go again—our knowledge that what we're doing is unhealthy is yet another stressor.

We need some different strategies to deal with the stresses of modern life. So—what if Ishmael was right, and Blue Mind is a better cure for what ails us? What if time spent in or around water was as effective as (and more immediate than) an antidepressant? What if we could treat stress, addiction, autism, PTSD, and other ills with surfing or fishing? What if your doctor handed you a prescription for stress or ill health that read, "Take two waves, a beach walk, and some flowing river, and call me in the morning"?[7]

Red Mind

In 2011, prior to the Blue Mind 2 conference, Sands Research conducted a study to examine the responses of a focus group of forty-five women who were viewing videos that included commercials, *Saturday Night Live* skits, and three clips of ocean environments: a turtle swimming, an undersea kelp forest, and a shot taken over the shoulder of a diver. They measured brain response via EEG and by monitoring eye movements to assess what Steve Sands calls *emotional valence:* the positive/negative, approach/withdrawal reaction to each video. By tracking the electrical activity in both hemispheres of the inferior frontal gyrus of the brain, reported Sands Research VP Brett Fitzgerald, "We can see the level of emotion people are experiencing moment to moment when they are watching a particular image."

The Sands EEG readings indicated that water images in the study evoked both positive emotions and sustained attention in the minds of the focus group. Interestingly, cognitive engagement was significantly lower when people were viewing the water videos. In fact, this indicated that the ocean images were a kind of "mental rest period" for the focus group—similar to the ways that many people experience a greater sense of rest and renewal when they are near water.[8]

Neuroscientist Catherine Franssen teamed up with World Surfing Reserves ambassador João De Macedo to further describe the differences between the two states I like to call Blue Mind and Red Mind, particularly how these two mental "maps" show up on or around water. Franssen, an expert on the biology of physical and mental stress, started off by further defining Red Mind as an "edgy high, characterized by stress, anxiety, fear, and maybe even a little bit of anger and despair." This state is a result of the physiological stress response that evolved to help us survive. "Your neuroendocrine system has been built and evolved for a reason," Franssen comments.

"These 'Red Mind' hormones are essential for escaping predators, and finding and fighting for food and mates."

That cocktail of norepinephrine, dopamine, and cortisol that flooded Jamal Yogis's system when he thought he saw a shark is exactly what occurs in a Red Mind state. It allowed him to escape a perceived predator by heightening his senses and giving him access to a burst of physical power he didn't have a moment earlier. Those Red Mind neurochemicals also can produce feelings of exhilaration and a hyperactive awareness that can be extremely useful in a range of extreme sports like skydiving, BASE jumping, rock climbing—and big-wave surfing. As João says, "We're not doing lotus poses in the impact zone. It's an extremely complicated area, with a lot of turbulence, and our wits need to be extremely sharp."[9]

"We need to have the stress response; it's important," Franssen adds. "But today, non-life-threatening stressors activate the same biological systems, meaning the same physiological stress response that we use to run away from a lion on the Serengeti is activated when the mortgage bill shows up in the mail. As we encounter little stressors throughout the day, our stress hormones remain high and keep us in an agitated place. Thus, our Red Mind stress response is turned on all the time, repeatedly, every day. Unfortunately, some of those stress-related neurochemicals, such as cortisol, can damage our bodies for up to two hours after even the smallest single stress." Repeated and sustained stress can wreak havoc from head to toe. In fact, the top ten causes of death around the world can either be caused or exacerbated by stress.[10] By sensitizing the amygdala through constant, inappropriate arousal, and weakening the hippocampus and preventing it from growing new neurons, increased stress can affect our ability to learn, retain information, or create new memories. Increased cortisol and glucocorticoid deplete the norepinephrine that helps you feel alert. Yet they also lower the production of dopamine and reduce serotonin levels, ultimately leaving you feeling flat, exhausted, and depressed.[11] Studies also have shown that the neural circuits responsible for conscious self-control are highly vulnerable to even mild stress.[12] "Repeatedly activating the stress response system is killing us," says Franssen.

So why are we not changing our ways?

Much Ado about Nothing

In the second decade of the twenty-first century it's almost impossible to conceive of a life without any digital component. When I'm not on the water or too isolated to get a signal, I will check my e-mail at least once a day—if not hourly. I don't have the fanciest phone (my kids remind me that I'm overdue for an upgrade and my screen is cracked), but what I've got can certainly do a lot more than just make calls. As you can see in the notes section for this book, my research has been made much easier by scientific journals and researchers posting their work online (often in open-access journals), so other scientists can have easy and immediate access to it. I've got a Twitter feed (though I can't claim my dispatches match the frequency and reach of the most tech-savvy scientists) and my own websites have proven incredibly useful for spreading the Blue Mind word. The challenge is figuring out how to balance, or at least harness, direct, and control the wave of technology at our fingers, because with supervision comes focus, occasional creativity, and a healthy compartmentalizing of our Red Mind energies.

According to former Microsoft Research Fellow Alex Soojung-Kim Pang, over the course of a typical day you will send and receive more than one hundred e-mail messages. That's an average of course, one that might be well below your daily total. But that's just the beginning: on that same day you'll "check your phone thirty-four times, visit Facebook five times, spend at least half an hour liking things and messaging friends... for every hour you spend talking to someone [on the phone], you spend five hours surfing the Web, checking email, texting, tweeting, and social networking." All of this adds up to an incredible 90 eight-hour days per year.[13] Pang also points out that recent studies have calculated that "a majority of workers have only three to fifteen minutes of uninterrupted working time a day, and they spend at least an hour a day—five full weeks a year—dealing with distractions and then getting back on task."[14]

Stop for a second and think about that: over the past year you spent more than *one month* dealing with distractions. Author Daniel Goleman has pointed out that "routine disruptions from a given focus at work can mean minutes lost to the original task. It can take ten or fifteen minutes to regain full focus."[15] That means it takes many more hours just to get the stuff done than if you could have stayed on track.

But it isn't just a question of adding hours to the day, as much as we try to do that in the hope of catching up. All-nighters, extra cups of coffee, weekends at the office—such measures can buy us minutes, but at a huge cost both to our physical well-being and our professional success. To see why, we need to look a bit more closely at what we call multitasking, which is really nothing of the sort.

Multitaskers Anonymous

Human beings have always been multitaskers. We can talk while we walk, we can sing in the shower, we can look toward the goal and shoot the puck, we can read out loud—and on and on and on. UCLA anthropologist Monica Smith believes that the development of such multitasking ability has been a crucial component of human evolution. "Multitasking in the modern era is at a whole new level of complexity, but it's really built on the basic skill sets that people already had," she says. That doesn't mean she thinks all is well, however. "The negative aspects of multitasking are much worse nowadays than they were in the past."[16] So what's changed to make everything worse? Let's think about what "multitasking" means today.

When was the last time you were driving around and saw someone in another car talking on their cell phone or texting, multitasking as they rolled along? Actually, says Paul Atchley, a cognitive psychologist at the University of Kansas in Lawrence, having a conversation with someone on the phone (or texting) while driving a car means performing two tasks at once, not one "multitask." Indeed, says Atchley, most people don't really multitask, they toggle back and forth between separate tasks.[17] In so doing, something has to give. When a conversation becomes more dense and complicated, cognitive demand devoted to it increases, and that means less brainpower is available for driving. As Pang explains: "*Multitasking* describes two different kinds of activities. Some are productive, intellectually engaging, and make us feel good. Others are unproductive, distractive, and make us feel stretched thin."[18] Wonder which type texting while driving is? A study done by the Virginia Tech Transportation Institute found that in 80 percent of car wrecks, the driver was distracted during the three seconds before the collision. As Peter Bregman, a columnist for the *Harvard Business Review,* puts it, "In other words, they lost focus—made a call, changed the station on the radio, took a bite of sandwich, checked a text—and didn't notice something changed in the world around them. Then they crashed."[19]

Remember a few chapters ago when we discussed how the brain has to filter out a great deal of incoming input in order to function? Even with all of our neurons, we can't deal with *everything*—our cognitive capacity simply isn't that great, as amazing as it is. Explains Atchley, when a conversation becomes more dense and complicated, cognitive demand devoted to it increases, and that means less brainpower for everything else.[20] You don't end up being able to handle multiple things at once, so you distribute your attention. Former Apple and Microsoft executive Linda Stone sums up the situation this way: "Today, we know that the brain processes serially—so quickly, that it may feel like we are doing two things at once, but we are actually just shifting very quickly.... Continuous partial attention is an always-on, anywhere, anytime, anyplace behavior that involves an artificial sense of constant crisis. We are always in high alert when we pay continuous partial attention. This artificial sense of constant crisis is more typical of continuous partial attention than it is of simple multi-tasking."[21] Or as Pang explains,

"During switch-tasking, your brain spends so much energy doing basic management that you have little bandwidth left for seeing previously invisible connections or making new associations."[22]

Superficially, this seems almost ridiculous. Isn't this the age of connection and association? Given how important some issues can be, isn't it a good thing that we have twenty-four-hour-a-day access to certain people and information? Unfortunately, we've cloud-seeded a deluge that threatens to wash us away. The Carnegie Mellon University sociologist, psychologist, political scientist, and economist Herbert Simon once said, "A wealth of information creates a poverty of attention."

The concept "the more we have to take in, the less we can handle" is supported by studies of the brain and our biochemistry. However, the harder our brain tries, the less information is actually absorbed. (If you've ever tried to cram for a test at the last minute, this reaction will be very familiar.) Among the scientists whose research has buttressed this conclusion is Michael Merzenich. When monkeys in his experiments performed mindlessly repetitive tasks, there was no lasting change in their brains. It was only when the monkeys paid close attention to the task at hand that strong new brain maps were established.[23] Our glucose-hungry brains, ever masters of efficiency, want to relegate as much as possible to autopilot to save energy, so that energy can be used on making maps of the new stuff. Trying to process multiple streams of information at the same time is no better for the brain. A 2009 study conducted at Stanford's Communication Between Humans and Interactive Media Lab showed that people considered high multitaskers (who regularly used a lot of media at one time) had more trouble paying attention, poorer memory control, and less ability to switch efficiently between tasks than the low multitaskers.[24] Overwhelmed, the high multitaskers failed to develop neural networks; their brains were like the flailing man at the bottom of a deep sandpit, scooping furiously only to find each portion of sand heaved upward was replaced by a comparable amount sliding back down, exhausted with not much to show for the effort. Peter Bregman acutely pointed out just how damaging such multitasking can be: "A study showed that people distracted by incoming email and phone calls saw a ten-point reduction in their IQ. What's the impact of a ten-point drop? The same as losing a night of sleep. More than twice the effect of smoking marijuana."[25]

We're getting the worst of all worlds: constant stress that taxes our bodies to the point of needing medical attention, and cognitive overwhelming that results in bad decisions. We lose focus, don't notice something changed in the world around us, then we crash.

Attention, Distraction, and How Water Can Reshape the Brain

> *. . . . neuroplasticity is activated by attention itself, not only by sensory input. Emotional arousal may also be a factor . . . the same factor may be involved in activating neuroplasticity when we participate in an activity that is important or meaningful to us.*
>
> — Barbara Bradley Hagerty, *Fingerprints of God*

The fact that most of the world now lives in urban settings makes the demands on our attention even worse. "When you walk down the street there are thousands of stimuli to stimulate your already overtaxed brain—hundreds of different people of different ages with different accents, different hair colors, different clothes, different ways of walking and gesturing, not to mention all the flashing advertisements, curbs to avoid tripping over, and automobiles running yellow lights as you try to cross at the intersection," writes Arizona State University professor of psychology Douglas T. Kenrick. "Research shows that in the absence of natural restoration, the human brain copes with this clutter by going into overdrive..."[26] And unfortunately, that mental overdrive creates the chronic stress that Catherine Franssen described earlier. Modern-day humans are drowning in a sea of overstimulation that exhausts us physically, mentally, and emotionally. To this scenario we often add coffee and sugar, powerfully caffeinated energy drinks, ever-brighter lights and screens, and even fast-paced music, creating an exhaustion factory—a downward Red Mind spiral.

Accessed in the proper way, Red Mind can actually help us learn to evaluate and reduce stressors in our lives more appropriately. An avid skydiver in earlier years, Franssen had noticed that jumping out of planes during the weekend helped her feel more peaceful, smooth, and calm during the week.[27] "I learned what was important and not to sweat the small stuff, whether it was a test or a paper or competition between my colleagues," she said. She theorized that exposure to the heightened hormonal tempest of Red Mind in extreme sports can actually help their participants recalibrate their stress response to daily life. So she did a study of extreme climbers, tracking their cortisol levels when they were exposed to a typical stressful situation (taking a test). As predicted, the stress levels of the climbers were lower than those of the control subjects. However, most of us aren't quite ready to "dial down" our stress response by bungee jumping or rock climbing. Could time in nature be an easier solution? Laura Parker Roerden, executive director of the Ocean Matters nonprofit, which immerses young people in the ocean through science and community service, explains, "Focused time in nature activates other parts of our brains, giving our fatigued frontal lobe [associated with executive function, cognitive control, and supervisory attention] a break. Areas of our brain associated with emotions, pleasure and empathy can now take over, providing a calming influence that is measurable in brain scans and blood tests alike."[28]

In the 1980s Stephen and Rachel Kaplan, two environmental psychologists from the University of Michigan, named this exhaustion "directed-attention fatigue." In a later paper,[29] they theorized that there are two kinds of attention: *directed*, which requires a great deal of energy and focus; and *involuntary*, which requires little to no mental effort. Directed attention is what you use when you focus on a task, make a decision, interact with others, pay attention to the road when you're driving, take a call, send a text, or choose what you want for dinner. In environments where we must use directed attention almost exclusively, our brains become fatigued, and our mental effectiveness is diminished. On the other hand, involuntary attention occurs when we're in an environment outside our ordinary habitat, with enough familiarity that it poses no threat but enough interest to keep the brain engaged. In this kind of environment, small things randomly attract our attention, but only momentarily, and usually require little focus or response—thus giving an exhausted brain a chance to replenish itself. The Kaplans theorized that when involuntary attention is highly engaged, direct attention can rest; and they believed that a natural setting was best for helping the brain to switch from directed to involuntary attention. "Our ancestors evolved in a nature-filled environment," stated Stephen Kaplan. "[Such places] *should* feel more comfortable, more relaxed, more like home."[30]

Over the years, the Kaplans and other scientists and psychologists have continued to do research on the restorative effects of exposure to natural environments. One 2008 study gave participants a test designed to fatigue their directed attention, and then had them take a fifty- or fifty-five-minute walk through either the university arboretum or the city's downtown. After the walk, participants took the same test. The scores for those who walked through the arboretum were significantly higher than for those who walked through downtown.[31] In 2013, researchers in architecture, environmental psychology, health studies, and urban design collaborated with Emotiv, a neuroengineering company that has developed a lightweight, multichannel wireless EEG that resembles a headset used by gamers. The researchers outfitted study participants with these headsets and had them take twenty-five-minute walks through three different areas of Edinburgh: a shop-lined street, a path through green space, and a street in a busy commercial district. The headsets sent back continuous recordings from five channels, indicating short-term excitement, frustration, engagement, arousal, and meditation (which, in technical terms, correlates to combinations of what neurologists call alpha, beta, theta, delta, and gamma brain waves). According to the study, when participants moved into the green-space area, frustration, engagement, and arousal dropped, while contemplation increased, and when they moved out of the green space, engagement rose.[32] (These results were similar to those in the Sands Research focus group study.)

What is it about the natural environment that provides such "rest" for the brain? Not long after his talk at Blue Mind 1, Michael Merzenich spoke about this from the neuroplasticity perspective.

> There are two parts to the process by which we create a model of the world. First, the brain is constantly trying to record and interpret the meaning of things and events, and it's continually changing itself as it constructs the model of the things in the world that will matter to it. But second, it's trying to control and indeed suppress the things that don't matter. You can think of it as normalizing the background.

In a natural environment, and in particular, when we are on or by the water, there is a high degree of statistical predictability because it is so much the same from moment to moment. The background is fairly controlled and a little dreamy—in other words, highly normalized, which allows part of the brain to relax. Against that background, the brain is continually looking for perturbation, for something that doesn't fit, something that moves, something that wasn't there before and that doesn't match with my reconstructed representation of the landscape. After all, the essence of survival in a landscape is the correct interpretation of the things that don't fit. But when the brain sees a perturbation, it creates a sense of surprise and novelty.

Such novelty is a crucial Blue Mind advantage. As John

Medina describes in his book *Brain Rules*, the quick-and-dirty assessment done by most animal brains regarding novel stimuli is: Can I eat it? Will it eat me? Can I mate with it? Will it mate with me? Have I seen it before? Michael Merzenich explained it in a conversation—from the point of view of a primate brain—following Blue Mind 1:

> Here's how it works in terms of neuroplasticity. Studies with macaque monkeys have shown that the brain very positively changes itself to more strongly represent the perturbation against the normalized background. All of the connections that would be engaged when they represent the thing of interest are strengthened, and all of the rest is weakened. So in a sense the brain is suppressing the background while it's heightening the perturbation in the moment. Now, natural landscapes are far less complicated than artificial human landscapes, and against a background that's less complicated, the brain can really see the perturbations. Therefore, in this sort of natural environment, any event that occurs—a bird landing on the shore, or a fish jumping out of the water, for example—is going to immediately captivate the brain's interest and attract its attention. In fact, the reaction to whatever it does see is heightened because the brain is able to focus more energy there. So the natural environment of being by the water is an ideal one for being engaged in a way that continually entices and intrigues the brain, and the brain responds to each perturbation with special interest and a wonderful little moment of neurological surprise.

And what, in the natural world, would be an ideal example of the ideal combination of background and perturbation?

Think about it: water is changing all the time, but it's also fundamentally familiar. It seems to entertain our brains nicely with novelty plus a soothing, regular background. Envisage yourself being by the water: the sounds, the sights, the smells, all changing moment to moment yet essentially staying the same. It's regularity without monotony—the perfect recipe to trigger restful involuntary attention. It's also the inverse of our current condition of monotonous suffocation.

Blue Mind... Take Me Away[33]

From the early Greek and Roman physicians who recognized the healing powers of nature and bathing; to stressed-out industrial workers in nineteenth-century England and the United States who were advised to "take the waters" by the seaside or at natural springs to recover; to men and women today who treat their drug and alcohol addictions or PTSD with the dopamine rush of surfing or with the endorphin serenity produced by long, calm hours with a fishing rod in their hands; to patients who stare at aquariums in dental waiting rooms and feel reduced anxiety after watching the fish; to the millions who step into a hot bath or shower at the end of a long day and emerge relaxed, refreshed, and renewed—all of these are examples of how water can help us transition from the Red Mind of stress or the Gray Mind of numbed-out depression to the healthier Blue Mind state of calm centeredness.

In examining the effects of water on the mind, there are three key questions to answer. First, how does being in or consuming water affect the brain? Second, how does being around watery environments support our mental health and well-being? And third, how do water, water sports, water views, water sounds, and so on help to heal our minds from the effects of chronic stress, PTSD, traumatic brain injury, depression, and addiction? In other words, how much truth is there in Hippocrates's observation that "water contributes much towards health"[34]? Only a few steps from your bedroom is a place where you'd be perfectly suited to find out. Indeed back in 1980 Calgon described its bath products as a cure for the stresses of a hectic life: "lose yourself... lift your spirits... take me away!"

Water, Water, Outside and In

> *There must be quite a few things that a hot bath won't cure, but I don't know many of them.*
>
> — Sylvia Plath

There are few humans who haven't experienced some version of their own "immersion therapy." The hot shower that wakes you up or unknots your body at the end of a long day; the hot-and-cold-contrast baths professional athletes use to ease strained muscles; the bliss of hot-tub jets pounding legs and arms and backs; or the delight of a long soak in a footbath or tub, perhaps filled with Epsom salts or essential oils designed to relax body and mind. As far back as the ancient Egyptian, Indian, and Roman civilizations, we have therapeutically immersed ourselves in water. The ancient Greeks, who viewed many diseases as being caused by spiritual or moral pollution, incorporated cleansing with water as a key element in healing rituals, and many of their temples, such as the one in Epidaurus, were situated near mineral springs. The Romans constructed spas in places like Bath, in England, where the water flows out of natural mineral springs at a constant temperature of at least 45 degrees Celsius, hence the location's appropriately simple name. In the Middle Ages and beyond, places such as Baden-Baden, Saint Moritz, Vichy, and Evian touted the physical and mental benefits of drinking and bathing in the local waters. Hot springs used for therapeutic bathing exist across the globe, notably in Australia, New Zealand, Russia, Canada, Brazil, Iceland, and the United States.

Spa bathing also significantly decreased levels of salivary cortisol (an indicator of stress) in college students,[35] and decreased both cortisol levels and self-reported mental fatigue in men who experienced "mild-stream bathing," where the lower half of the body is continually massaged by a stream of water.[36] Back in 1984, psychologist Bruce A. Levine studied the effects of hot-tub immersion on fourteen patients diagnosed with anxiety disorders. To eliminate the placebo effect (in which people improve simply because they are told that an experiment is designed to treat their condition), patients were told that the session was designed only to measure their anxiety and they should expect no reduction in symptoms. After fifteen minutes of hydrotherapy there was a significant reduction of subjective anxiety, as well as a decrease in muscular tension as measured by electromyography (EMG) levels.[37] A study at around the same time showed that anxiety levels in test subjects dropped after a five-minute hot shower.[38] Several studies have shown that immersion in warm water during the early stages of labor helps lower anxiety, decrease cortisol levels, and promote relaxation in women.[39] In a randomized and controlled study of the effects of hydrotherapy and three other treatments on 139 people with rheumatoid arthritis, the hydrotherapy patients not only had greater improvements in joint tenderness and in range of movement, but also exhibited improvements to their emotional and psychological states.[40] And as a sign of therapeutic potency, in Japan, where bathing is a social as well as a personal ritual, several contemporary studies have shown that hot-water immersion can increase activity in the parasympathetic nervous system while decreasing it in the sympathetic nervous system (an indication of relaxation)—the extent of such immersion sometimes no more than a footbath.[41]

When considering the mental health benefits of water, however, it's equally important to examine how the *lack* of water (or indeed, too much water) affects the workings of the brain. Remember, the human body is 60 to 78 percent water, and the brain is even more "watery" (up to 80 percent by volume). It's no great surprise that consuming enough water is a requirement of healthy brain function. Even mild dehydration can affect the brain structures responsible for attention, psychomotor and regulatory functions, as well as thought, memory, and perception,[42] and has been shown to decrease reaction times in working memory, lower alertness and concentration, and increase fatigue and anxiety in adults.[43] Even scarier, there is some evidence from studies done in rats that cognitive impairment due to dehydration may not be completely reversible due to cellular-level damage.[44] As we age, dehydration becomes an even greater risk, and, according to geriatric medicine specialists Margaret-Mary Wilson and John Morley, has been shown to be "a reliable predictor of increasing frailty, deteriorating mental performance, and poor quality of life."[45] In children, the effects of dehydration on cognitive function are equally serious; this is especially concerning as some studies indicate that two-thirds of children in the United States, the United Kingdom, Italy, and Israel are dehydrated when they arrive at school.[46] Luckily, giving children additional water can improve performance on tasks involving visual memory, verbal recall, processing speed, and reaction times.[47] (Of course, too much water—or too much water consumed too quickly—can cause problems by upsetting the electrolyte balance in the body, causing the cells of the brain and body to swell and produce headaches, confusion, drowsiness, even changes in behavior. However, for most adults "too much water" would mean drinking more than 5 liters over two hours; most adults in the U.S. consume only around 2.5 cups of water a day.[48]) It's one thing to be psychologically in need of some Blue Mind, but if you're physically parched, you can't do much of anything at all.

How Being Around Water Improves Mental Health

> *...no person could be really in a state of secure and permanent health without spending at least six weeks by the sea every year. The sea air and sea bathing together were nearly infallible...*
>
> —Jane Austen, *Sanditon*

In 1660 in England, a man known as Dr. Witte published a now obscure book titled *Scarborough Spa*, in which he described drinking and bathing in seawater as a "Most Sovereign remedy against Hypochondriack Melancholy and Windness."[49] In the 1700s and 1800s, Georgian and Victorian physicians prescribed visits to the seaside or spas as restorative for both physical and mental health. (It didn't hurt that from 1789 to 1805, King George III visited the seaside town of Weymouth, ostensibly to relax and help cure his bouts of madness.)[50] Today, more than three and a half centuries later, a group of psychologists, medical researchers, and physicians in the United Kingdom are studying the effects of water proximity on physical and mental health, and one of them, Professor Michael Depledge, has been at the forefront of what I call the blue health movement.

Mike Depledge combines the professional and passionate assuredness of an acknowledged world-class expert with the composure and authoritative delivery of a BBC nature documentary host. In 2009, he and physician William Bird of the Peninsula Medical School in Plymouth announced the establishment of a program called the Blue Gym.[51] "Nature helps to stimulate us to be more active," Depledge asserted in a video announcement of the program. "Coastal areas, beaches, and inland waterways appear to be particularly efficacious in this regard. Regular contact with these and other natural environments reduce health inequalities by providing three major health benefits: a reduction in stress, increased physical activity, and stronger communities.... The Blue Gym will help to connect even more people to the sea, rivers, canals, and lakes."[52]

In the years since then, Depledge and other researchers at the European Centre for Environment and Human Health have been exploring the effects of blue nature on health. In 2012, along with his fellow researchers Benedict Wheeler, Mat White, and Will Stahl-Timmins, Depledge analyzed 2001 census data for England's 48.2 million residents. They looked at how many people who lived within fifty kilometers of a coast or estuary reported having "good" (as opposed to "fairly good" or "not good") health as opposed to those who lived farther from water. The portion of those saying their health was good was 1.13 percent higher in communities less than fifty kilometers from a coast. While this is a relatively small increase, the health benefit of living closer to a coast rose as the socioeconomic status of respondents dropped; in other words, living near water may help to ease some of the typically negative effects on health caused by lower socioeconomic status. "Controlling for both individual (e.g. employment status) and area (e.g. green space) level factors," the research team wrote, "individuals reported significantly better general health and lower levels of mental distress when living nearer the coast."[53]

Where does this greater self-reported health come from? It's possible that the simple fact that people can walk to, or readily visit, the water makes the difference. A 2013 paper tracked the activities and attitudes about the sea of fifteen families with young children (ages eight to eleven). "Although families valued the opportunities for physical activity and active play afforded by beaches, the key health benefits emphasized were psychological, including experiencing fun, stress relief and engagement with nature," wrote study authors Katherine Joan Ashbullby, Sabine Pahl, Paul Webley, and Mat White. "Increased social and family interaction were also highlighted as benefits."[54]

Still, heightened physical activity was likely part of the equation, too. A study of Australians in New South Wales showed that those who lived in a postal code on the coast were

27 percent more likely than the rest of the population to report activity levels that were adequate for healthy adults (higher than the "sedentary" category), and 38 percent more likely to report vigorous levels of activity than inland dwellers.[55] Back in England, "As over two-thirds of all coastal visits were found to be made by people who live within five miles of the coast," Wheeler and his colleagues reported, "coastal communities may attain better physical health due to the stress-reducing value of greater leisure time spent near the sea."[56] In a review of eleven scientific studies done of physical activity outdoors and indoors, researchers found that "exercising in natural environments was associated with greater feelings of revitalization and positive engagement, decreases in tension, confusion, anger, and depression, and increased energy."[57] More important, a separate review of studies of the benefits of "green exercise" indicated that exercise in proximity to water enhanced both self-esteem and mood more than green space alone.[58]

Water Heals the Mind

Kayaking is my medicine.

— Ed Sabir, Vietnam veteran

The *Heroes on the Water* video starts quietly, with an image of sparkling, golden-lit water, and on it, silhouetted, someone in a kayak, fishing rod in hand. A few minor-key notes from a piano play in the background as the picture switches to another man, in shorts, shot from the waist down, standing next to a row of kayaks lined up on a riverbank. Before you register that the man has only one leg, the picture switches again, to an overhead shot of someone sitting in a bright yellow kayak, paddle in one hand... and no arm below the elbow on his other side. A voiceover says, "Life's not over after an IED." IEDs (improvised explosive devices) were used extensively against U.S.-led coalition forces in Iraq and Afghanistan, and were responsible for well over 60 percent of coalition deaths.

Next you see three young men in their twenties, one in a wheelchair, all missing limbs. One of them—buzz-cut hair, T-shirt, with a prosthetic left leg, black and shiny, visible below white cargo shorts—looks at the camera and continues with the kind of confidence one exudes when imagining vast possibilities ahead: "If you can fish, hell, what else can you do?"[59]

Since 2007 Heroes on the Water (HOW) has helped more than three thousand wounded warriors and veterans to relax, rehabilitate, and reintegrate by taking them kayak fishing. According to Jim Dolan, HOW's founder, kayak fishing provides triple therapy: physical (because the vets are paddling and fishing), occupational (as they learn new skills and a sport they can practice for life), and mental (due to the freedom and relaxation that accompanies being on the water). "I know when I'm out there on the water all the crap in my life goes away," Dolan says, "and so I figure it's the same for them."

Dr. Jordan Grafman has studied the functions of the human brain for more than thirty years. As the former chief of cognitive neuroscience at NIH's National Institute of Neurological Disorders and Stroke, and currently as the director of brain injury research at the Rehabilitation Institute of Chicago, he is a leading expert in TBI and the effects of TBI on PTSD. Grafman's enthusiasm for his area of expertise belies the serious nature of the injuries he helps treat, and unfortunately, over the past decade or more he's had a lot of people to study. "In civilian populations, between three to six percent of people tend to be eligible for a diagnosis of PTSD," he reported to attendees of the Blue Mind 3 conference. "In victims of natural disasters, like earthquakes or tsunamis, it's between four and sixteen percent."[60] However, as many as 58 percent of people who are exposed to combat are diagnosed with PTSD. According to a 2008 RAND Corporation report, more than 620,000 men and women who served in Iraq and Afghanistan since 2003 returned with PTSD, major depression, or TBI.[61]

PTSD is the result of the brain's dysfunctional response to a traumatic experience (or experiences), or the ongoing wear and tear of multiple stressful situations, or prolonged, persistent grief over the loss of friends and loved ones. "Someone is exposed to a traumatic event and responded with fear, helplessness, or horror," Grafman says. "They lose their sense of personal safety or the feeling of being master of their environment. They may also experience shame or survivor guilt. Then they reexperience the event, either in similar situations, or in dreams or imagination, and become hyperaroused." Every repetition of the event strengthens the connections between the event and the painful feelings, and tends to make people (often irrationally) fear that the same event will happen again in the future. Here we see another example of how the brain's plasticity can be a curse, a neural map drawn and redrawn in Red Mind ink. "Many people with PTSD can't sleep," reports Bryan Flores, a member of the Monterey County Mental Health Commission who has fought his own battles with seasonal affective disorder (SAD). "Every time you close your eyes, you see that car crash or whatever caused the traumatic stress." Good sleep is critical to good dreaming and good dreaming is fundamental to creativity, learning, and memory.[62]

In PTSD, as with all stress, the parts of the brain most directly affected are the amygdala, hippocampus, and medial prefrontal cortex. "The amygdala is concerned with emotions and fear, but also the relevance of what we see in the world to our personal safety," Grafman continues. "The hippocampus is important for forming and retrieving everyday memories, which the amygdala labels with emotional content. The medial prefrontal cortex is important for storing beliefs, routine memories, and episodes we retrieve all the time. Both the amygdala and medial prefrontal cortex seem to be most important in producing PTSD."

So how might programs like Heroes on the Water help? Speculates Grafman, "Sometimes people just want to surrender to the soothing, pulsating rhythms of waves. Being by water both relaxes and stimulates us and therefore our brains, and it leads to positive emotional changes in human behavior. I suspect that brain activity in many of the same brain regions implicated in post-traumatic stress by the direct experience of water could help to calm people and enable them to produce an adaptive response to PTSD." He adds that "multiple studies of brain activity show that the medial prefrontal cortex is quite active during moments of insight, but just as important, it is affected by a positive mood. The more positive mood someone has, the greater insight, and the more activity in that area of the brain. Helping someone to feel more positive moods is going to support brain activity in the medial prefrontal cortex."

John Hart, a neurologist who is the medical science director of the Center for Brain Health in Dallas, and who works with the veterans at Heroes on the Water, says something similar: "Water impacts all five senses at the same time with a very positive, powerful image and memory. The good memories from their day on the water help override the bad memories and images that haunt them, and possibly help crack the shell, letting them rejoin the world." And water isn't passive: in most instances (and certainly in cases involving a kayak and a fishing rod) we have to interact with it. As Grafman notes, "We're both in awe and curious about it, and it's a challenge to master." It distracts us in the best sort of way, allowing us to think of little else beyond what's in front of us.

It's not just that fishing produces a sense of calm and the release of endorphins associated with sustained pleasure; getting out on the water or into wilderness settings can be deeply restorative in itself, according to the branch of psychology known as ecotherapy. Being in the wilderness can produce feelings of respect and wonder, a greater sense of connection with oneself and with nature, of renewal and greater self-awareness. "Wilderness therapy" has been used with psychiatric patients, emotionally disturbed children and adolescents, survivors of sexual abuse, cancer patients, addicts, and AIDS patients, as well as individuals with PTSD.[63] Indeed, programs like HOW, Project Healing Waters Fly Fishing, Rivers of Recovery (fly-fishing), and Casting for Recovery (fly-fishing for women recovering from breast cancer) provide what for many is a rare chance to participate in activities in beautiful, healing, calming locations. "[These programs] let Mother Nature do what she's done for hundreds of years, and that's to heal," says Butch Newell of HOW.

Rivers of Recovery (ROR) takes veterans on four-day trips

to rivers in wilderness locations around the country, teaching fishing skills and relaxation techniques. In 2009 researchers from the University of Southern Maine, the University of Utah, and the Salt Lake City Veterans Administration studied sixty-seven veterans on three occasions: one month before, the last day of, and one month after an ROR trip. To assess physiological stress, the researchers measured salivary cortisol, urinary catecholamines, and salivary immunoglobulins. They also asked the veterans to self-report their mood, depression, anxiety, and perceptions of stress. There was significant improvement on every measure following the ROR trip. A month later, veterans reported that their PTSD symptoms and perceptual stress had gone down 19 percent; physical symptoms of stress decreased by 28 percent; sleep quality improved 11 percent; depression lessened 44 percent and anxiety 31 percent. Feelings of serenity increased 67 percent, assuredness by 33 percent, and positive mood went up 47 percent.[64]

These clinical results are borne out by anecdotal evidence. One young man, Eric, had been blown into the side of a helicopter while he was getting ready to parachute, cracking his helmet and putting a half-dollar-size hole in his head. His chute then failed to fully deploy, so he fell a thousand feet and landed on his head. He survived with injuries to both frontal lobes, no short-term memory, and panic attacks caused by severe PTSD. After Eric and his wife participated in an ROR trip, his wife reported that he hadn't had a panic attack in almost a year. Another veteran, James, who returned to ROR as a volunteer, said that he had been able to eliminate all of his medication for depression and most of his pain medication after his time on the river. Jason, a vet who participated in

HOW kayak fishing trips wrote, "I had a world convinced I was okay, but my demons won more battles than I did. I know kayak fishing saved my life."

Other water sports also have been turned into therapy for veterans and civilians alike. Programs like Operation Surf (run by my buddy Van Curaza); the National Veterans Summer Sports Clinics, which offer sailing, surfing, and kayaking every summer; AmpSurf in Central California; and Ocean Therapy sessions offered to members of the Wounded Warrior Battalions at Camp Pendleton and 29 Palms, California, are designed to help those who have been injured in some way to overcome their challenges and heal through participation in water sports. As AmpSurf's mission statement says:

> One in five Americans struggle with a life-long disability. . . . Whether they are an amputee, visually impaired, suffer from PTS (Post Traumatic Stress), or have quadriplegia . . . whether they are a child with autism, or a young woman who has lost a limb to cancer, AmpSurf offers a unique program to bring the healing power of the ocean and adaptive surfing together for an experience that is both mentally and physically one of the best forms of rehabilitation on the planet. Our goal is simple, we want to help our disabled veterans, adults and children focus on their abilities not their disabilities.

A while ago I hit the waves with two British vets, Martin Pollock and Will Hanvey, who had come to Santa Cruz from the United Kingdom for one of Van's Operation Surf camps (more on that session later). Martin had lost both legs and one arm in the wars in the Middle East, and Will was injured during an intense training exercise when heavy downward pressure was put on his knees, dislocating and severely injuring them and his hips. Like most camp participants, they had arrived thinking that there was no way they would ever be able to surf. But Van tells camp participants, "Look, you're not going to surf like me; you're going to surf like you. We'll take your physical abilities and apply them to this surfboard." Watching these vets in the water is something else. There's a sense of achievement that destroys their perceived limitations, lifts their self-esteem, brings them together with others like themselves, and rebuilds their psychological strength.

On top of that, veteran and 2014 National Geographic Explorer of the Year Stacy Bare told a group of us at a meeting convened at the Marines' Memorial Club in San Francisco by the Sierra Club's Mission Outdoors that "many veterans feel like the most interesting, exciting, and important experiences of their lives are behind them." Feeling that there is no longer any need for (or physical ability to carry out) the suite of high-octane, adrenaline-rich, highly technical skills they've acquired can exacerbate detachment, lower esteem, and lead to depression and addiction. But Martin and Will found their Operation Surf sessions so transformative that they returned home and started Operation Amped U.K. in Cornwall.

A Texan named Bobby Lane came to Operation Surf to cross surfing off his bucket list; after returning home from the camp he planned to kill himself. Lane, twenty-seven years old, suffered a traumatic brain injury while serving in the U.S. Marine Corps in Iraq. Back home, like so many others he battled PTSD and couldn't sleep due to incessant nightmares. He says Operation Surf not only changed his life, it saved his life. Surfing also allowed him to get his first good night's sleep in years. Dreams about water replaced his midnight terrors, Blue Mind at its most potent. "Giving up on life isn't an option," he says. "Now, as I see it, if life gets too hard, there's always the ocean."

In August 2013 Van Curaza asked the question "What is your favorite thing about the beach?" on his popular Facebook page. One person commented "the waves," another "the sun." The usual responses appeared as one scrolled down: "the air," "the surfer dudes," "the sound," "the salt," "the ocean," "our kids." Way down the list there's Bobby Lane's stirring comment: "The peace it brings to my life. When things get hectic in my life, or when things start to get bad again. All I have to do is go to the beach, and everything that is going on just seems not to matter anymore. So my favorite thing about the beach is the peace it brings me."

Water Heals: Addictions

The waves of the sea help me get back to me.

—Jill Davis

Van originally realized the transformative power of surfing when it helped him overcome his addiction to drugs and alcohol. And he's not alone. Jaimal Yogis tells the story of Jamie Patrick, who kicked a cocaine habit cold turkey when he discovered triathlons and ultra swimming.[65] Today programs like the one run by pro surfer and recovering addict Darryl Virostko (also known as "Flea"—hence the name of his program, FleaHab) in Central California and Surfing to Recovery in Southern California use outdoor activities to help "replace the high of drugs with the endorphins of exercise."[66]

In such cases, healing on the water actually comes from replacing the rush produced by addiction and other at-risk behaviors with the more natural dopamine "high" produced by surfing, white-water kayaking, sailing, or competitive paddleboard racing. "The goal of surf therapy is not to teach people to be surfers," says mental health commissioner Bryan Flores. "It's to get them to use surfing to change their brain chemistry. You stand on the beach and get amped up and all kinds of chemicals rush through the brain. Different ones are in play when you're paddling out or have a monster wave chasing you to the beach. All of those chemicals can have incredible effects on how people cope with depression, anxiety, stress, and other mental health issues."

According to neurobiology studies, addiction is a reward prioritization problem that involves long-term neuroplastic changes in the brain. "Addiction is where people start exploiting just one mechanism of reward to the exclusion of other possibilities," states Vanderbilt's David Zald. "Drug-related rewards cause direct changes in the system—the same system that makes us seek more natural rewards like food or sex—and make us more and more exclusively linked towards the addictive drug." Again and again we've seen evidence that addiction involves disordered dopamine systems concerned with reward and motivation. These alterations draw addicts strongly and selectively to the reward of the drug and overwhelm any sense of risk or consequences. Howard Fields calls this "an inflated value signal": "Dopamine neurons and the nucleus accumbens (the 'pleasure' center of the brain) are activated by certain action outcomes, such as the use of an illicit drug, and the inflated value of the drug. Consequently, in future decisions, actions leading to drug use are selected at the expense of all other activities." However, as we've seen, neurobiology is not destiny, and a brain that has changed in one way has the potential to change in another. To change an addictive behavior, "the person has to be able to identify a stronger alternative reward... to overcome the compulsion to seek or engage in the addictive behavior," notes Zald.[67]

Surfing and other sports provide such alternative rewards by satisfying the brain's desire for stimulation, novelty, and a neurochemical "rush," while also getting addicts out of their typical environments (a critical aspect of most recovery programs)[68] and providing new settings, new friends, and new routines that are as fulfilling but more positive. States Fields, "'Positive' addictions that result in individual and/or societal gain are thus desirable behaviors—such as an 'addiction' to nature or being outdoors, which stimulates positive feelings in response to immersing oneself in the natural environment."

In the summer of 2010, Kevin Sousa was lying in the guest room of his house in Manhattan Beach, California, stoned out of his mind. He had taken his first drink at age thirteen, been in and out of rehab for more than a dozen years, and at age forty realized that it was time to do something. "I am addicted to anything and everything that makes me feel good/bad, and I am tired of it," he wrote in his journal. "I have never quit or accomplished anything in my life of true merit. So today, I start training for the great Catalina to Manhattan Beach Paddleboard race. I have been on a paddleboard once or twice, but I figure I need to trade one addiction for another." For almost a year Kevin trained with a friend, paddling for hours in the ocean four days a week. He used AA, his "new addiction" of paddleboarding, and the creative act of writing music about his journey as the means of his recovery. On August 28, 2011, the day he completed the Catalina Classic paddleboard race, Kevin wrote, "Doing inspiring and healthy things has a curious way of spreading. I have never felt quite so alive and competent as a man and a human being." Kevin has maintained his drug-free life while his wife has battled cancer, and he has become a fine therapist and an inspiration to others—all because, as he says, he chose to "return to the ocean, a place that has brought me immeasurable joy."

Addiction is a complicated illness, and it would be irresponsible to claim that going Blue Mind is an inescapable means to purity. But for an increasing number of those struggling with the most merciless of addictions, the curative effects of water are impossible to deny. And if water can make such a difference in combating their ferocious demons, what might it do for you?

Autism and Water

Going into the water you leave autism behind...

—Don King, whose autistic son, Beau, bodysurfs with him

While neuroscience and cognitive psychology have made great strides in understanding why being in or near water is so good for mental and physical health, there are some watery "miracles" that are yet to be explained. One of these is the positive effect of water and water-based exercise on children with autism. Clinicians who study aquatic programs designed for autistic kids discovered that parents and recreational therapists report increases not just in swimming skills, muscle strength, and balance in the children, but also greater tolerance of touch and ability to initiate/maintain eye contact.[69] In a Taiwanese study, children had fewer negative behaviors, and demonstrated greater attention and focus and more appropriate conversations with peers following a ten-week swimming program.[70]

But it's on the beach, in the waves, that some of the biggest transformations take place. In Deerfield Beach, Florida, a fit, white-haired, fiftyish surfer dude named Don Ryan has been running a program called Surfers for Autism since 2007. Professional surfers and other volunteers come together with around two hundred autistic kids and their families for a day of surfing, paddleboarding, music, and games. Children with many of the classic traits of autism—including lack of ability to focus, limited communication skills and ability to verbalize, anxiety around others, lack of social skills and inappropriate responses, difficulty forming personal attachments, a sense that they are in their own world, and extreme sensory sensitivity to light, sound, smell, repetitive movements and ritualistic behavior—often approach the water with fear and trepidation. But time and time again, once they're in the water, or on a surfboard or paddleboard, something good happens. Kids who have rarely smiled or spoken wear wide grins as they ride a wave to shore. They come out of themselves and start to relate to people around them. Dave Rossman, who volunteers with the group, comments, "Once they are on the beach, you can't tell a kid with autism from any other child." [71] One mother at a Surfing for Autism event said, "I will never forget the joy I saw on my daughter's face that day, the pride she felt that she was enough just the way she was." At these events, her daughter "isn't a girl with Asperger's—she's just a girl who is seven, catching a wave."[72]

There are all kinds of theories about why this happens: the water is stimulating visually, which fulfills some children's sensory needs; water provides "a safe and supported environment" that surrounds the body with "hydrostatic pressure" that "soothes and calms"[73] (as another expert said, it feels like the "ultimate hug"[74]). Learning new motor skills like swimming, surfing, or paddleboarding can have "a broad-ranging impact on the nervous system," according to William Greenough at the Beckman Institute at the University of Illinois: "There's increased blood flow to crucial neurons, and the reshaping of abnormal structures in the front brain. But beyond that, surfing may be a vehicle to an emotional breakthrough, a way of reaching under the mask and perhaps connecting to kids like these." Trying to balance and ride waves also provides them with a clear focus and keeps them in the present moment—neurobiologist Peter Vanderklish believes that the beauty of surfing "turns the focus of these kids inside out. They're pulled out of themselves by having to live in the moment, and all their anxieties are pushed aside."

When he was all of thirteen years old, Naoki Higashida, who suffers from autism, wrote a remarkable memoir that was later translated into English and published under the title *The Reason I Jump*. In it, he lyrically speaks about the effect of water on himself and others who are autistic:

> In the water it's so quiet and I'm so free and happy there. Nobody hassles us in the water, and it's as if we've got all the time in the world. Whether we stay in one place or we are swimming about, when we are in the water we can

> really be at one with the pulse of time. Outside of the water there's always too much stimulation for our eyes and our ears and it's impossible for us to guess how long one second is or how long an hour takes.... We are outside the normal flow of time, we can't express ourselves and our bodies are hurtling us through life. If only we could go back to that distant, distant watery past—then we'd be able to live as contentedly and as freely as you lot![75]

In 2012 former pro surfer turned Patagonia surf ambassador and documentary filmmaker Keith Malloy produced a film about bodysurfing that featured Don King of Kailua, Hawaii, and his autistic son, Beau. The pair were diving through the crystal waters off of Oahu, riding the waves together. "No one smiles like Beau underwater," Keith commented. On the other side of the Pacific Ocean, in a coastal town in New South Wales, Australia, another couple with an autistic son were at their wits' end. Donna and Greg Edwards's eight-year-old son, Kyan, had kept them from going to the beach for over two years—"The bright sunlight, sand and wind create a 'sensory overload'...and most often he will scream and scream when we take him anywhere near the ocean," Greg wrote in a letter to Keith.[76] "My wife and I both grew up surfing and we feel completely at home in the water. Some days, the two of us have wanted nothing more than to be able to take our kids down onto the sand and swim." One day Greg, Donna, and their two boys watched Keith's documentary, and Kyan couldn't keep his eyes off of Beau King. After the program, the boy started running around the house with his arm outstretched, imitating the posture of the bodysurfers he had seen. Then he ran toward his mother, saying,

"Beach...surfing." The very next day the family went to the beach, and Kyan went straight into the water. "That day, and every day since, have been remarkable," Greg wrote. "Our son has discovered the ocean. He now feels at home in the water. His confidence is growing and he is catching his first waves!"

"You deal with autism on such a raw basis every single moment of every day," Donna added. "To be in the water and kind of leave that behind—it's huge." However, Michael Rosenthal, a pediatric neuropsychologist of the Child Mind Institute, who works with autistic children, comments, "As a clinician I'm always working with kids to find something that's going to make them feel happy and safe and secure and relieve their anxiety. But it's important to note that water in the ocean is therapeutic for a lot of people. I don't think it's necessarily an 'autistic' thing as much as it is a human thing."[77]

I've sat with Jack O'Neill—wetsuit developer, businessman, and philanthropist—over recent years, in his home overlooking Pleasure Point in Santa Cruz, talking about Blue Mind and sharing personal stories. Many of the uses of Jack's "Wetsuit Medicine"—from surfing to open-water swimming, paddle-sports to scuba diving—are described in the pages of *Blue Mind*. Like many I've met, Jack has felt the healing power of H_2O.

Jack has made it a point to be in the water on the average of once a day for the last fifty years: "The surf has been my life and livelihood, and the ocean has been a great healer," he commented. "I'm just a surfer who wanted to build something that would allow me to surf longer because being in and near the ocean was and is my medicine. It can fix what's broken. Even today, at age ninety-one, I jump in the ocean and everything becomes all right."

"Take Two Waves..."

Take a course in good water and air; and in the eternal youth of Nature you may renew your own.

—John Muir

In May 2012 I was driving our bright orange Kubota tractor with a brush-cutting attachment on back—a pretty big rig, weighing in at about a ton. I was clearing an old fire road we like to hike and run up that winds up Schoolhouse Ridge above Mill Creek. A stroll in the forest is good for mind and body, but a dash through the redwoods by water makes my neurons glow blue. With the spring rains, the plants had grown in like a jungle, with lots of nettles and spiny brush. I was zipping along the road as it climbed the hillside above the creek.

What happened next took a split second but felt like slow motion. (I discovered later that this is a neurological phenomenon called *tachypsychia*. It's believed to be caused by high levels of dopamine and norepinephrine, and it alters our perception of time in moments of great physical or psychological stress.) The tractor rode up on a big log that had fallen in a recent fire and was hidden under the plant growth. When I bounced off the log, the tractor lurched toward the edge of the bank, which immediately gave way, causing the tractor, with me on it, to roll down the thirty-foot cliff. There was no time to think: it happened in the time it takes you to snap your fingers.

I remember arms, legs, orange steel, flashes of blue sky and green trees, then stillness. I was on my back in Mill Creek, looking up, a smashed, upside-down tractor next to me, still running. My first thought was "Am I dead?" I blinked hard, said "Hello" out loud, then, "Am I very broken or bleeding?" I felt my head and neck: all okay. I checked for blood: some, but not gushing. Next thought: my family. They don't even know I was on the tractor today. They're fine; they're not worried because they don't know I'm here under a tractor!

The next thought: "Is there a leak from the tractor into the creek?" I pulled myself out and up, found the tractor key, and shut it down. I needed rags, sponges, towels—something to make sure none of the oil from the tractor entered the water. I followed the creek downstream until it ran along a dirt road, climbed up the bank, and then started running the three-quarters of a mile toward home. I was benefiting from the upside of Red Mind, the physiological "fight or flight" storm of neurochemicals—adrenaline, cortisol, and norepinephrine—coursing through me. My broken bones, dislocated shoulder, burned skin, rattled brain, torn ligaments were nonexistent. My brain chemistry and concern for the endangered salmon that call Mill Creek home overrode any possibility of reason.

When my partner, Dana, came home she found me, naked and in a state of shock, sitting in the redwood hot tub on our deck. (My first instinct had been to get in the water.) She took me to the hospital, and on the way we recruited my brother and our neighbors (who, unlike me, are experienced ranchers) to bring towels, bales of straw, and a big skidder and winch. They extracted the tractor from its precarious creek-bottom location among the redwoods and alders without harming the sensitive habitat.

At the hospital, under the gaze of doctors, nurses, and X-ray machines, the pain set in, bruises appeared, and every joint swelled and locked. I had a mild concussion, broken toes, torn foot ligaments, a hyperextended hip, a previously dislocated shoulder that had popped out and back in again, and bruises down to the bone on both legs. My brain had come down from its Red Mind high, and I felt exhausted and depleted. (The effects of an intense adrenaline/dopamine/norepinephrine dump linger in the bloodstream, and the letdown is dramatic.) For a month afterward I found myself in the throes of Gray Mind, a numbed, depressive state mixed with acute jitters. I kept seeing and reliving the accident, replaying those three seconds like a movie, micromoment by micromoment, and obsessing over all the ways I might have avoided going over the cliff. The doctor diagnosed post-traumatic stress; all I knew was that my brain and body were both a mess. I was an all-round, lifelong athlete with excellent balance and coordination, yet now I couldn't walk down steep stairs or descend an escalator without the feeling of rolling off a cliff. I had been an impeccable speller, but now I was making mistakes on the simplest words.

The months since the accident were marked by a slow recovery, aided by water. I took long soaks in the bathtub with tons of Epsom salts at first, then later I could get into the hot tub, my girls beside me to lift my spirits. I hobbled around on crutches for a month and a cane for a couple of months after that. I went swimming (well, bobbing) in the cold Pacific—therapy for both my banged-up body and mind. One day I walked back to the place where the tractor and I had landed. Standing in the icy waters of Mill Creek at the base of the cliff also helped to heal the mental trauma.

Soon after the accident I embedded myself with Van Curaza's team at Operation Surf. It didn't take long before I was in the water, side by side with guys like Martin Pollock who had lost limbs in the Middle East. Of course, I had all *my* limbs, a clear advantage in the water, but a malfunctioning, rattled brain doesn't care what did the rattling—tractor, IED, car crash, falling off a cliff or down a flight of stairs. The guys accepted me into their motley band of surfers, but each morning out on the waves I felt frustrated and weak. My balance was shot, and forward motion brought up that vertiginous feeling of rolling, falling, crashing, and a need to grab a rail. But still, we were all in the water, having fun, cheering each other on. Then Van said, "This one's yours, J. Don't think, no mind, eyes on the beach." I paddled hard, hopped up, and rode it. No thoughts. No rolling. No falling.

That wave felt like a never-ending exhale, the opposite of a micromoment, a time warp, jazz, a painless birth, summer in my soul, catharsis, a lifetime somehow packed into a now. I looked over my shoulder and there was Martin surfing the same wave, all the way to the beach. Blue Minded and fully stoked, we smiled at each other and paddled back out.

The bath when I got home was pretty nice, too.

7

Blue Unified: Connection and Water

Empty your mind, be formless, shapeless—like water. Be water, my friend.

— Bruce Lee

On December 31, 1999, the last day of the twentieth century, in the U.S. Virgin Islands, actor Michael J. Fox had an encounter with a sea turtle that changed the direction of his life.

> As the turtle and I swam together in disjointed tandem, the turtle trying to ignore me and me trying to pose no threat, I thought of all those documentaries I had watched as a kid: thousands of hatchling baby sea turtles making their way toward the safety of the ocean while seabirds dive-bomb, picking them off one by one. Only a handful will survive. And that's just the beginning of a turtle's ordeal. I noticed that this one was missing a sizable chunk of the rear flipper on his left side. How old was this guy? I wondered. An adult, obviously. What wars had he been through?[1]

Something in that encounter caused Fox to walk out of the water, up to his wife, Tracy, who was sitting on the beach, and tell her he was leaving his hit TV show, *Spin City*, at the end of the season. It seems that "all my epiphanies come to me at or near a beach," he wrote later.

Fox had revealed to the world in 1999 that he suffers from Parkinson's disease, and after he left *Spin City* in May 2000, he launched the Michael J. Fox Foundation for Parkinson's Research. Since then Fox has been a tireless fundraiser and advocate in the search for a cure for Parkinson's. And when he returned to television, he did so with a different sense of why—not as a career, but as an experiment that fit into a more holistic sense of self: "[Your challenge] doesn't have to be life-shattering or life-ending.... It can just be a new thing that pushes you to a new place. And so like when I thought about doing this show, it's like, why not? Why can't I? Why can't I?... You don't have to shut it down. You don't have to withdraw. You don't have to pull in."[2] In other words: you don't have to stay on the shore once you know you're standing on a shore.

I believe I know how Michael J. Fox felt on that beach. Having spent my life around water and its creatures, I can testify that something mystical can happen to the mind and heart when we intersect with nature. Humans are surrounded by man-made buildings, objects, and environments, and it can become harder and harder to remember our intimate relationship with this beautiful blue planet. But magic can happen in the fleeting moments in which we notice the natural world—the sunset that causes us to catch our breath, the murmur of wind rustling through trees, the sharp, clean smell of rain on grass or the tang of salty air near a shore, the feel of sand or dirt underfoot. These moments reconnect us not only to nature, but also to our *own* nature; they carry with them the recognition that we are part of something bigger than ourselves. As psychologist William James wrote, "We are like islands in the sea, separate on the surface but connected in the deep."

Humans have always sought to understand our relationship with what lies beyond us. At times, that "beyond" has been the natural world. At other times, "beyond" has been what we believe created the world and us, or a sense of the interconnectedness of everything. The feeling of connection is called by many names: empathy, compassion, awe, transcendence, ecstasy, love, wonder, enlightenment, flow, unity, mindfulness. It is felt during prayer, meditation, and worship; in tasks that take us out of ourselves, and in the moments when we care for others; on days and evenings when we marvel at the rising sun or the stars. We feel connection in settings from the humblest of temples to the most beautiful synagogues, mosques, and cathedrals; in forests or mountains or lakeshores; when we stand next to sleeping children or gaze into the eyes of loved ones, or when we see or touch the creatures of this world and marvel at the magnificent diversity of life. James called this connection mystical consciousness, a form of awareness entirely different from our typical waking state. "No account of the universe in its totality can be final which leaves these other forms of consciousness quite disregarded," he said.[3]

Many, if not most, people greet the terms "meditation" and "mindfulness" with an eye roll or the profession that "I don't do that stuff." This isn't surprising: modern culture puts a premium on motion, on multitasking, and on increasing urbanity. We've so glamorized "science" and what's seen as the "Western" mindset that the idea of sitting still in hope of transcendence seems comical, its practitioners worth mocking. But in recent decades, the scientific perspective on meditation has drastically evolved. For comparison, it wouldn't be a stretch to say that just as over the past half-century we have realized that quitting smoking has enormous health benefits and chronic stress plays a key role in many diseases and disorders, so too have we begun to conclusively realize that meditation offers tremendous advantages—and we've realized this by using those "Western" filters. What science is also revealing is that there's an additional simple, watery means to mindfulness. Indeed, think of it as Blue Mindfulness.

Skip This Chapter (and Get Back to Work)

It's amazing how many companies these days promote some kind of meditation program—and not just in Silicon Valley. Target, Procter & Gamble, General Mills, Comcast, BASF, Bose, and New Balance are among an increasing number of corporations to offer mindfulness training and encourage its use at work.[4] They're doing this not only because they care about their employees but—let's be honest—because they care about profits. "Mindfulness is an idea whose time has come," says Google's Chade-Meng Tan. "For a long time practitioners knew, but the science wasn't there. Now the science has caught up."[5]

Play Misty for Me[6]

"Mindfulness is the process of actively noticing new things," explains Harvard University psychologist Ellen Langer. "When you do that, it puts you in the present. It makes you more sensitive to context and perspective. It's the essence of engagement."[7] Alex Pang echoes this belief: "The popular image of meditation is that it's a kind of blissed-out blank-mindedness," he says. "Nothing could be further from the truth."[8] In fact, studies show that our mind wanders and blanks out not during meditation but during stress and switch-tasking. Daniel Goleman offers a striking example of such blank-minded meandering:

> A reader's mind typically wanders anywhere from 20 to 40 percent of the time while perusing a text. The cost for students, not surprisingly, is that the more wandering, the worse their comprehension.
>
> Even when our minds are not wandering, if the text turns to gibberish—like *We must make some circus for the money,* instead of *We must make some money for the circus*—about 30 percent of the time readers continue reading along for a significant stretch (an average of seventeen words) before catching it.[9]

Replace "perusing a text" with "driving" and "gibberish" with "a deer running in front of the car" and you're looking at a hard slap in the face by an exploding airbag, if not worse. But while less violent than a collision, imagine the productivity impact of having to constantly reverse course because you belatedly realize things aren't adding up correctly. Two steps forward, one step back still makes for progress—but wouldn't you prefer three steps forward?

Think about such stuttering, inefficient advancement in relation to a 2011 study done by researchers at the Princeton University Neuroscience Institute using fMRI and other measurement tools to evaluate the brain's response to a disorganized environment. They found that physical clutter distracts the brain, creates greater stress, overloads the senses, and makes it more difficult to process information. Other studies confirm that our brains are not built to function optimally in such a distracting and cluttered mental (or physical[10]) environment. But in our distracted, overwhelmed world, we're not only encountering environmental clutter via our own distraction, but *making it*. So, to take Goleman's example further, we receive hasty e-mails that (given the sender's distraction) have mistakes, and (due to our own distraction) skip over the errors when we read them, compounding the problems. (I wish for you to be the recipient of the first e-mail I send after getting out of the water—it is undoubtedly the most clear and concise one I'll send all day.)

Of course, distraction is not new to the twenty-first century. Indeed, one of the first major studies of its effects was after the Second World War, when researchers were interested in why the targeting accuracy of radio operators declined as their shifts went on.[11] And we've known forever that the more tired we are, the more mistakes we make. But how many of us have a job (running a grocery store, raising kids, teaching high school students, tuning skis, overseeing mergers and acquisitions) that allows us to tell the boss and our colleagues we're going to head home for a good nap and then return to finish up? How many companies are willing to establish "z-mail" policies that shut off access to employees' e-mail accounts over the weekend?[12] (That would be next to none.) And yet without cognitive refreshment, can anyone expect the work that we do to be as good as it should be?

In the previous chapter we saw how water can help drug addicts kick their habit. But just as a little Red Mind can be like a little red wine—perfectly wonderful at the right time and in the right quantity—too much Red Mind leaves us in a near-permanent state of guilty dependence and digital delirium tremens when we try to step back even for a moment. In *Fast Company* magazine, Mark Wilson sympathetically describes our addiction, in this case focusing on our phones, but the argument extends much more broadly to our distance from the natural world.

> Claiming that it's our fault when we can't disconnect is like saying it's our fault that we can't eat just one Lay's potato chip, smoke just one cigarette, take just one hit of heroin, or have just one orgasm. Addictions are irresistible by nature—by our core physiology, psychology, and everything else that makes us fleshy, fallible humans. And since when do we, as reasonable members of society, blame the addict who's been hooked on a substance engineered by a giant, publicly traded corporation? Maybe it's time we look beyond branding and become a bit more honest about what the iPhone has become: An enticingly useful, often unsatisfying, instinctually craveable drug. At the very least, maybe it's time we all stop blaming the victim.[13]

But just as those drug addicts found a form of salvation in

the waves, we can find it right in front of us if we make just a bit of effort.

The Creative Advantage

Imagination is more important than knowledge. Knowledge is limited. Imagination encircles the world.

— Albert Einstein

Creativity is a big subject, important in *everything*—from art to business to science—and in everyone—from young children to nonagenarians. So it's not surprising that scientists have been trying to discover the neurological "maps" of creativity, to see if it's possible for humans to train ourselves to access that creative state more often. "Creativity is a renewable resource, one that's universally, if not evenly, distributed," wrote *Time* science and technology editor Jeffrey Kluge. "We don't decide how much we get, but it's up to each of us... to tap what's there."[14]

Creativity on command can be a challenge for many of us, but it's the bread and butter of jazz musicians and improv comics, among others. In 2012, when researchers at the National Institutes of Health wanted to examine the brains of people in "creative flow" (which we'll talk about later in this chapter), they found twelve experienced freestyle rappers, put them in fMRI scanners, and asked them to (1) improvise lyrics to an eight-bar musical track and then (2) perform a memorized set of lyrics to the same track. The scans revealed that the areas of the brain associated with emotion, motivation and initiative, language, and motor skills were all more active in freestyle rap than with the rehearsed lyrics. At the beginning of improvisation, the left hemisphere was more active, but as the riff ended, the right hemisphere was the more engaged. Freestyle rap, which is an amalgamation of language and music, seems to activate areas involved with both. During the rap, activity in parts of the frontal lobe (notably the dorsolateral prefrontal cortex) associated with executive function (processes involved in planning, organizing, strategizing, paying attention, remembering details, and managing) decreased—perhaps indicating a shift from directed attention to looser, more "uncensored" activity. This was "consistent with the notion that a state of defocused attention enables the generation of novel, unexpected associations that underlie spontaneous creative activity," the study authors wrote.[15] In a complex world where almost everyone has access to the same wealth of information creators may not be aware of where their inspiration comes from, since the "top-down" attention of the conscious mind goes quiet during bursts of improvisational creativity. However, that doesn't mean the conscious mind has no role to play in creativity. Just as focused attention is needed when we experience flow while surfing or kayaking, when the brain moves to the second phase of creation—refining and revising—executive functions have a significant role to play.[16]

As a scientist and an author, I've experienced both the rush of creative inspiration and improvisation as well as the detailed focus of revision, and it's pretty clear that water can help with both.

Back in 1951 psychologist D. W. Winnicott talked about creativity occurring in "the space between the inner and outer worlds."[17] If, as novelist Alexandra Enders wrote, creativity "incorporates memory, imagination, intention, and curiosity, but also exists in the real world,"[18] finding the space and conditions that help an artist to access that creative state is vital. In Santa Cruz, not far from where I live, there is a small inn called Ocean Echo right on the beach. It's simple, nothing to write home about; certainly the rooms are utilitarian at best, except for the presence of the ocean immediately outside, steps away. The gentleman who owns it tells me that he has customers, musicians and writers, who regularly come down from San Francisco or up from Los Angeles to stay on the beach for a week. For them, the chance to be away from their usual surroundings and by the water—water they can hear, see, and smell—clears the cobwebs out of their brains and gets them back into the creative state. "Most of us... have a special place or two that brings out special dimensions of ourselves," writes Winifred Gallagher. "Profiles of artists invariably include a paragraph about the settings and rituals they depend on to help summon up and sustain their creative states."[19] Such places provide a framework, an anchor, an entry point into creativity. And perhaps because water is all about flow, many artists find these places by oceans, lakes, rivers, streams, and (famously, in Thoreau's case) ponds.

Thus I don't think it's an accident that so many artists' colonies grew up in places with water nearby—Provincetown on Cape Cod; in Montauk, New York, in Giverny, France, and Cornwall, United Kingdom. Benjamin Franklin, Edmund Rostand, and Vladimir Nabokov did much of their writing while in bathtubs.[20] Author and Booker Prize–winner Hilary Mantel reports that she takes showers when she gets stuck in her writing.[21] Oliver Sacks reportedly got over his writer's block with long swims every day in Long Island Sound:[22] "There is something about being in water and swimming which alters my mood, gets my thoughts going, as nothing else can," he writes. "Theories and stories would construct themselves in my mind as I swam to and fro, or round and round Lake Jeff. Sentences and paragraphs would write themselves in my mind, and at such times I would have to come to shore every so often to discharge them."[23] Robinson Jeffers built his own home, Tor House, stone by stone, in Carmel, California. Pablo Neruda lived by the ocean in Chile on and off for thirty-three years, and wrote that in Isla Negra he attended "the university of the waves." In 2014 artist and filmmaker Julian Schnabel told the BBC's Philip Dodd on the radio program *Free Thinking* that "the freedom of surfing...an otherness...the reason people are drawn to surfing, the reason they really do it is because they can get to a place they can't get to any other way. There's great liberation in that. If you look at different surfers over the years...they have a style the same way painters have a style and actors have a style or poets have a style.... It's not really a sport... it's a way of life."[24]

It is the layers of rhythmic, structured symphony performed by waves and wavelets, stones and pebbles. It is the known shallows that taper into the mysterious abyss. Water is both lover and mother, murderer and life-giver, source and sink. It is the endless mutability, the surprise and unexpectedness of its ever-changing colors and moods that stir artists, musicians, writers, filmmakers, and thinkers alike. Water unleashes the uninhibited child in all of us, unlocking our creativity and curiosity.

The Watery Muse

> *The ocean stirs the heart, inspires the imagination and brings eternal joy to the soul.*
>
> — Wyland

A year or so ago I was visiting my cousin Julia in my native New York City with my daughter Grayce, and we went out to Red Hook, a heavily industrial area of Brooklyn next to the "other" bridge to the borough, the Manhattan Bridge. We walked the shoreline back toward the Brooklyn Bridge to visit Ran Ortner's studio to see his huge ocean canvases, some of them eight feet tall and thirty-two feet wide. All you see in Ran's paintings is ocean. There are no bridges, landmarks, boats, people, animals; just ocean waves rendered with such detail and richness that at first you could think you were seeing a photograph instead of an oil painting. I thought it a little ironic that Ran's studio had no views of any kind, and certainly not of creativity-inspiring water. But when Grayce, Julia, and I walked in the door, it felt as if Ran had re-created the ocean in all its majesty. His paintings have the same impact as the ocean itself. Some people spontaneously well up and cry. A feeling of awe is common, as is the desire to spend some time with his work. No wonder there's a big comfy couch right there where you need it in the center of the room!

When you ask Ran what his art is about, he'll tell you, "I paint the ocean. Period. Why? Because I love it. Period." If you can get him to talk about it a little more, he'll say things like, "What you see is exactly how I experience the ocean. Out of that ancient body comes this pulsing energy, like a metronome constantly marking the Now." Or, "I thought how powerful it would be if I could bring to painting even a fraction of the immediacy I feel when engaging with the ocean." He says that he wants his paintings to have an impact so immediate that they become "what Kafka called an 'ax for the frozen sea inside us.' "[25] You may begin to understand why kicking back on that couch at Ran's place might be so emotionally and intellectually appealing.

Many of the other artists whose works I have seen or read about have in common their desire to use water as a means of communicating a vision and connecting with others in a visceral way. In the past several years neuroscientists have started to study the neurological basis of our response to art and to beauty—a field called neuroaesthetics. Researchers have discovered that the neurological processes that occur when we perceive something aesthetically involve brain structures responsible for perception, reward, decision making, and emotion.[26] If you survey the world of art, however, it's pretty clear that "beauty" and "art" are subjective experiences encoded by culture and personal history. "A need to experience beauty may be universal, but the manifestation of what constitutes beauty certainly is not," write Bevil Conway and Alexander Rehding in their 2013 survey of neuroaesthetics. "Beauty...varies in complex ways with exposure, context, attention, and rest—as do most perceptual responses."[27]

While the scientists who study neuroaesthetics are still unclear about what exactly creates the experience of beauty and art in the brain, most of them will agree that there are three components: perceptual, cognitive, and affective (or emotional). The perceptual processes are the same as when we take in any stimulus (visual, auditory, and so on). The cognitive processes involve making sense of the stimuli and giving them meaning. But whenever I speak with artists, it's clear that what matters most to them is the emotional effect of their art.[28]

My friend Halsey Burgund assembles audio collages of music and voices speaking about the ocean. Choreographer Jodi Lomask creates a multimedia performance, Okeanos, at the Aquarium of the Bay in San Francisco, combining live dance, aerial performers, video projections of underwater dance and ocean elements, and interactive talks with experts to demonstrate "the spectacular vitality found in the sea." Writers from Homer to Melville to Twain to Conrad, poets from Coleridge to Mary Oliver to Lisa Starr use water as background and matrix for their stories and poems, pulling readers deep into the experience of ocean and river. Sculptors like Theo Jansen create kinetic Strandbeests, which move along the beaches of Holland, propelled by wind and weather. Photographers like Karen Glaser, Brian Skerry, and Neil Ever Osborne capture an instant of time and light in their images of water and the animals and people in and on it. All of them are trying to communicate even the smallest portion of the essence of their experience of water.

Of course, artists are inspired by different things. Some go to Santa Fe for the sky, some go to New York for the energy, some go to Algiers for the sensory overload, some go to Paris for the mood. But water's infinite variety and (sometimes terrifying) depth has an unrivaled inspirational force when it comes to the physical world. Its potential is metaphysical to the nth degree, and here, too, creativity and problem solving are naturally compatible. After all, what is creativity but a form of optimism that there is more that can be done?

Artsy-Fartsy and Touchy-Feely[29]

The creative act is a letting down of the net of human imagination into the ocean of chaos on which we are suspended, and the attempt to bring out of it ideas.

— Terence McKenna, author and philosopher

My sister Jill Hoy and her late husband, Harvard art professor Jon Imber, spent half of the year by the sea on Deer Isle, Maine. They painted together by the water for a long time, side by side for countless thousands of hours. They met on the island more than twenty years ago. Jill grew up visiting Deer Isle and painting the colors of the sea (our father owned a ship captain's house in the village), and Jon visited the island in summer to paint. Jill and Jon experienced water through different eyes. They saw lines and colors, shapes and patterns, on its surface and in its depths that we didn't know were there until they opened our eyes.

People fall in love with the light, the coast, and the ocean in the same way Jon and Jill did each time they painted. People sometimes want to remember that feeling. Sometimes they buy a painting so they can take some of that feeling back with them to put on a wall, to look again and again at the colors and the light held by the paint Jon or Jill placed on the canvas at the sea. Somehow, each deliberate stroke of a brush loosely held by a hand at the end of an arm, wired to a brain responding to the light reflecting off the water and passing through two eyes, is able to hold within it the magic of the ocean and the passion of the painter, and touch the heart of a viewer many miles away and many years from now.[30]

Perhaps because water is used in the creation stories of dozens of religions, or perhaps because water is so malleable, changeable, shimmering, yet with invisible depths, it is used frequently as a metaphor for the creative process. As University of Vermont professor of environmental studies Patricia Stokowski points out, bodies of water—oceans, streams, rivers, pools, fountains—are used both realistically and symbolically, as "a dominant feature in all forms of artistry and artistic expression."[31] Water can represent death, birth, destruction, play, a bottomless tomb or a familiar and beloved haven. In psychology, water is used as a metaphor for the unconscious and for emotion. (Psychologists may not discuss water in their pages, but take a look at the number of psychology books that feature it on their covers.) A passage on water can symbolize a heroic quest, the movement of time, a journey to the underworld, the education of the young, and the voyage home to the old. "Water has a nearly unlimited ability to carry metaphors," is "the fluid that drenches the inner and outer spaces of the imagination," writes Austrian philosopher Ivan Illich. "What it says reflects the fashions of the age; what it seems to reveal and betray hides the stuff that lies beneath."[32]

Metaphors and analogies are crucial for our understanding of the world and how we should interact with it. Douglas Hofstadter, a professor of cognitive and computer science at Indiana University, and Emmanuel Sander, a psychologist at the University of Paris, have written that analogies,

> far from being unusual cognitive gems, are mundane events, being generated several times every second, and it is through them that we manage to orient ourselves in the world. Analogical thought involves the perception of important but

> often hidden commonalities between two mental structures, one already existing in our brain, representing some aspect of our past experience stored in an organized fashion, and the other one freshly constructed, representing a new circumstance in our lives. In essence, an apt analogy allows a person to treat something new as if it were familiar. If one is willing to let go of surface attributes and to focus on shared properties, one can take advantage of past knowledge to deal with things never seen before.[33]

Language, we know, has its limits, and metaphor and analogy help us understand one another despite some of these constraints. This is especially true when we face new situations, since putting a label on anything necessarily involves ignoring some details in favor of an abstract analogue.[34] David A. Havas, director of the Laboratory for Language and Emotion at the University of Wisconsin-Whitewater, and James Matheson of the School of Cognitive Sciences at Hampshire College have noted that "Language can impact emotion, even when it makes no reference to emotion states. For example, reading sentences with positive meanings ('The water park is refreshing on the hot summer day') induces patterns of facial feedback congruent with the sentence emotionality (smiling), whereas sentences with negative meanings induce a frown."[35]

Abstraction can have a powerful cognitive effect well beyond communication. A few years ago, researchers explored how impressions and decisions might be influenced by "incidental haptic sensations"—that is, when we hold certain objects. The impact of our metaphorical foundation was unmistakable:

> In six experiments, holding heavy or light clipboards, solving rough or smooth puzzles, and touching hard or soft objects nonconsciously influenced impressions and decisions formed about unrelated people and situations. Among other effects, heavy objects made job candidates appear more important, rough objects made social interactions appear more difficult, and hard objects increased rigidity in negotiations.[36]

The opposite of rough and hard? Liquid H_2O.

Think about it: simply *touching* a hard object increased mental and psychological inflexibility. What might that mean when it comes to thinking creatively and finding new solutions? In some cases, getting to yes requires getting to wet.

But wait, I know what you might be thinking: *Yeah, I'm going to be trying to sell something to a customer, and when things get ugly I'm going to invite him to go hot tubbing?* (I'd say, "Sure!," but of course I live in Northern California, so I would, right?) Probably not. But how often are you in a position where you're interacting with another person and if you don't solve the problem right then, right there, all is lost? Emergency medicine, military engagement—not many where lives are on the line. There are many instances where immediate cooperation is essential: a restaurant kitchen, for example. But to some extent those interactions are very different from, say, trying to explain to your daughter why she can't spend her allowance on an Xbox. The problem is that we treat all such exchanges as Red Mind defined, just as we do with our e-mail, our voice mail, our spreadsheet all-nighters, Black Friday sales, and so forth. The greater the presumed stakes, the more important it is to be at our best. Think about one of your greatest negotiation disappointments. Was it as urgent as it seemed at the time? Would the result have been any worse had you told the other party you wanted to pause the conversation and resume later? Would the result have been any better if during that break you had sat in the tub, or swum some laps, or sat on a bridge listening to the sounds of the river below? I bet the odds of progress would have improved—"progress" meaning not just that a deal would be negotiated, but also that if an agreement was not to be reached (and sometimes that is the best outcome), you would be at greater peace.

I probably don't need to point out that an Xbox can be the least of it. Arms control agreements, buying another company, settling labor disputes—the way in which your creativity is encouraged and your rigidity decreased can make an enormous difference. So what *is* the marginal benefit of that trip to the swimming pool? Or the cost of avoiding it?

I Agree (with Myself)

Human beings are incredibly good at finding evidence to support whatever they believe and discounting any fact or theory that goes against that belief. Facts and figures do little to change our minds, because whenever we are shown facts that are contradictory to our opinions, the emotional circuits of the brain light up, not the reasoning ones.[37] Such rigidity is guaranteed to lead to bad decisions and to greatly diminish creativity. Daniel Goleman argues that "new value arises from the original synthesis, from putting ideas together in novel ways, and from smart questions that open up untapped potential."[38] Confirmation bias obstructs such inquiry, because it discards input that contradicts already settled conclusions.

But what if we could apply the power of analogy and metaphor to push back, to spark our creativity by peeling away confirmation bias's strict filter? One of the metaphors most associated with creative thinking is "fluid." "Such language reflects a metaphor for thinking about creative thought," write Michael Slepian of Tufts University and Nalini Ambady of Stanford University. "For instance, creative thought is often contrasted with analytical thought, which is more rigid and precise; a fluid can move in multiple directions with ease, and the ability to fluently and flexibly generate multiple thoughts is essential for creativity."[39] In a mesmerizing experiment, Slepian and Ambady had a group of volunteers trace two similar images, the difference being that one was made up of curves and flowing lines, designed to induce fluid arm movements, while the other was made up of straight lines requiring a more inflexible technique. (Imagine drawing a circle and then a hexagon—such was the test, albeit with more complicated images.) After completing their respective tracings—the participants had been told only that the experiment was to study hand–eye coordination—the volunteers were asked to come up with as many creative uses for a newspaper as possible within sixty seconds. Those who had done the fluid tracings came up with substantially more uses for the paper during that minute. Slepian and Ambady then went further, replicating the experiment with different tests, such as asking participants to complete math problems; those who had done the fluid tracings scored significantly higher than those who had gone the straight-line route on all of these.[40]

In their book *Mental Leaps: Analogy in Creative Thought,*

cognitive psychologists Keith Holyoak and Paul Thagard demonstrate the powerful applications of analogy to problem solving, decision making, explanation, and communication. Not surprisingly, waves are a recurring theme. "The concept of wave has developed from a specific analog tied to a particular kind of example, water waves, to an abstract category that can be applied to a vast range of situations involving the rhythmic propagation of patterns," the authors write.[41] The curved lines we make with our pencil bring to mind (literally) some of the deep reactions offered by actual waves. It seems incredible, but it's really just amazing, Blue Mind at work.

Hard and rough leads to psychological inflexibility and a decline in creativity. Straight and firm diminishes creativity and intellectual performance. Propagating ripples permeate art and science alike. These are glimmers of a greater illumination, one mapped into our brains over thousands of generations. Indeed, it shows just how finely tuned our neurochemicals are when even these faintly analogous properties and actions have such pronounced impact. These reifications, their prompts tactile, graphic, or semantic, are raindrops in the desert of unoriginality.

Channeling

As we've discussed, being on, in, around, or near water can calm our overactive minds while it imbues our senses. It does this by tapping into ancient neural maps and their associated neurochemical reactions. It can also help us access the state not coincidentally called by another watery word, "flow," allowing us to access the default-mode network/daydreaming parts of our brains while restoring our ability to focus and perform cognitive and creative tasks with greater ease. All the while it inspires some of the greatest art of the past as well as many of the cutting-edge artworks of the modern age.

And much of this optimum brain activity can start with a simple glance out the window or a walk by the shore.

The Fishbowl Effect (Outside Looking In)

We've talked before about various studies that have shown how even indirect exposure to water has recuperative power. Views of nature—whether through windows or in artworks—have been shown to help hospital patients feel better and recover faster.[42] Several studies of the effect of hospital gardens have shown that patients who take advantage of those spaces, be it by visiting them or even having a window view from their room, experience significantly less emotional distress and physical pain.[43] In one fascinating survey, people recovering from heart surgery looked at one of three scenes shown on panels at the foot of their beds. One scene showed an enclosed forest, another a view of open water, and a third an abstract design or blank white panel. As is consistent with other findings, patients looking at nature scenes needed less pain medication; what is especially interesting, however, is that the anxiety levels of patients viewing the open-water scene were significantly lower than for those looking at the enclosed forest.[44,45]

Viewing water and fish in aquariums also has been shown

to help lower stress and promote a better mood. A study done at the National Marine Aquarium in Plymouth, England, monitored the blood pressure, heart rate, and self-reported relaxation levels and moods of 112 people who spent a minimum of ten minutes observing an aquarium tank with three different levels of biodiversity (no fish or crustaceans; a few specimens; and a healthy variety of marine life). In all three conditions, blood pressure dropped substantially during the first five minutes in front of the tank, while the most positive changes in heart rate, relaxation, and mood occurred with the greatest amount of biodiversity. [46] Thirty years ago, in 1984, researchers from the Schools of Dental Medicine and Veterinary Medicine at the University of Pennsylvania performed a study of different treatments to reduce anxiety in patients prior to elective oral surgery: viewing an aquarium or viewing a poster, with or without a hypnotic induction. Viewing the aquarium, with or without hypnosis, proved significantly more relaxing; in fact, further analysis of data showed that hypnosis added nothing to the high level of relaxation experienced when people viewed the aquarium.[47]

Of all the things we've discussed so far, these results are among the most important. Yet despite all the positive data embracing a new lifestyle can be difficult. Asking someone to sit by the shore when they might be fired for doing so is asking a lot. But unlike all the other means of reaching mindful clarity, water can *do the work for you*. Not all of it, but some of it. To explain this further, we need to finally get to that one sense we didn't discuss earlier: sound.

The Hush and the Roar

> *The rain I am in is not like the rain of cities. It fills the wood with an immense and confused sound. It covers the flat roof of the cabin and its porch with insistent and controlled rhythms.... It will talk as long as it wants, the rain. As long as it talks I am going to listen.*
>
> —Thomas Merton, *Raids on the Unspeakable*

For years architects and urban planners have known that using water elements in city environments creates a better quality of urban life. "Urban landscapes are saturated with signals that carry little or no intentional information and are regarded as unwanted noise by many people. These signals emanate from vehicles (e.g., motors and road noise) and stationary machines (e.g., air conditioners)," write the authors of "Soundscape Ecology: The Science of Sound in the Landscape."[48] Sound pressure is measured on a logarithmic scale with 0 decibels being the lower range of what we can hear and the threshold for pain usually given at 140 dB, lower in children. An increase of 10 dB doubles perceived loudness and represents a tenfold increase in sound level. Thus 20 dBA would be perceived as twice as loud as 10 dBA, 30 dBA would be perceived as four times louder than 10 dBA, 40 dBA would be perceived as eight times louder than 10 dBA, and so on. Levels greater than 165 dB experienced for even a moment can cause permanent damage to the inner ear. In New York City maximum noise levels on subway platforms measured 106 dB and 112 dB inside subway cars.[49] To put that in perspective, the human voice in conversational speech is around

60 dB, and rustling leaves, a babbling brook, or water sliding along the hull of a kayak comes in at 20 dB.

You can see why our sound processing systems, built for millions of years to be sensitive to the smallest noises in the environment, are overwhelmed, exhausted, and sick. People who live in places with continuously high levels of traffic noise have a greater risk of high blood pressure, heart attack, and suppressed immune systems.[50] All of this noise isn't just annoying—it's lethal.

As a Harvard Medical School Senior Research Fellow specializing in the effects of sound, Shelley Batts is supremely qualified to speak about sound and water. "We spend our first nine months underwater, hearing sound through water in the womb," she comments. "We hear the whooshing of our mother's heart, her breath going in and out, the gurgle of her digestion.... These fluid, rhythmic sounds are very much like the ocean. Perhaps that's why the ocean often brings up feelings of relaxation and tranquility." From the womb onward, sounds have a profound effect on us physically, cognitively, and emotionally. And while prolonged exposure to loud noise causes the release of stress hormones and can lead to long-term damage, not just to our hearing but also to our general health,[51] pleasant sounds at comfortable levels have been shown to improve mood, induce relaxation, and enhance concentration.[52] According to Batts, sounds like that of water are inherently pleasant to our ears because they are not high frequency or harsh and feature a regular wave pattern, harmonic pitch, and low volume.[53]

"From a psychological perspective," Michael Stocker, an acoustician and author of *Hear Where We Are: Sound, Ecology, and Sense of Place*, told me, "the sound of water means life."

And sound is nearly always present. "If we occlude sound from our thoughts, it's only meaning we lose—the sounds that convey the message still strike our ears, resonate in our chests, and glance off our faces."

Petr Janata, from the Center for Mind and Brain at the University of California, Davis, is a cognitive neuroscientist and expert on music and the brain. He theorizes that the low frequency of the sound of water, coupled with its rhythmic nature, is similar to the frequency and rhythm of human breath. Sound, Janata contends, "affects our brain and influences our emotions. If I ask you to close your eyes and turn on a recording of the ocean, I can change your mood immediately." The sound of water evokes some of the same sensations as meditation, and studies by Japanese researchers show that the sound of a creek in the forest produces changes in blood flow in the brain that indicate relaxation.[54] Water sounds have been used by millions of people to help them sleep,[55] and ocean sounds can be remarkably effective in calming fearful patients. In 2012 a group of dentists in Malaysia played the sounds of water fountains for patients between the ages of twelve and sixteen prior to dental care. They found that the natural water sounds reduced the teenagers' worry and anxiety about treatment by nine percentage points compared to the control group.[56]

"In the brain, the medial prefrontal cortex is associated with linking sensory input (like sound) to subjective cognitive experience, emotional response, and self-reflection," Batts explains. "It's the same brain region responsible for feelings of compassion and connection. So it's very possible that pleasant sounds become linked easily to feelings of positive emotion and connection to other humans and our environment."

Other studies have elaborated on this association. Technically, "waves breaking on a beach and vehicles moving on a freeway can produce similar auditory spectral and temporal characteristics, perceived as a constant roar." As subjects were being scanned by fMRI machines, they were shown movies of freeway traffic (which researchers considered "non-tranquil") and waves crashing onto the sand. The results:

> Compared with scenes experienced as non-tranquil, we found that subjectively tranquil scenes were associated with significantly greater effective connectivity between the auditory cortex and medial prefrontal cortex, a region implicated in the evaluation of mental states. Similarly enhanced connectivity was also observed between the auditory cortex and posterior cingulate gyrus, temporoparietal cortex and thalamus.[57]

So under conditions of identical auditory input, ocean imagery resulted in significantly improved brain connectivity. Same sound, profoundly different results.

In 1997, a California researcher in psychoneuroimmunology (the study of the interactions between our nervous and immune systems and their relationships to mental health) showed ten cancer patients who were experiencing chronic pain a nature video that included fifteen minutes of the sounds of ocean waves, waterfalls, and splashing creeks. After viewing the video, patients experienced a 20 to 30 percent reduction in stress hormones such as epinephrine and cortisol.[58]

This is the huge advantage of water: you don't need to meditate to take advantage of its healing effects because it meditates *you*. And the listening doesn't need to involve sitting by sand castles or in a bobbing canoe, just as getting a visual Blue Mind boost doesn't require a visit to the shore or stroking between lane lines. Recall those hospital ward studies we mentioned earlier: just being able to *see* nature had incredible benefits, and we know by now that the best sort of nature to see has water in it.

Nobody's going to tell you a fountain in the middle of a shopping mall can do as much for you, Blue Mind-wise, as standing in a rushing current, fishing pole in hand, or body surfing the waves on a sunny afternoon. (Though such fountains have been shown to have remarkable effects.)[59] But what studies like these show us is that even something as modest as a fishbowl on your desk or a tiny fountain or the right kind of sound machine can work some real magic.[60,61] On my daughter Julia's desk is a medium-sized tank that is home to a snail and two small freshwater fish. Occasionally a visitor from the creek in our backyard is introduced for a brief visit (before being returned to its natural habitat). They have more than enough places to hide and an attentive keeper who feeds them religiously and eyeballs them endlessly, studying each move of the fins. I watch Julia watching the fish watch each other. Yes, keeping fish in a tank is not ideal for the fish, but I'm reminded of my own childhood explorations of water and the wonderful animals living in it. I can see Julia's curiosity and empathy grow with her technical knowledge. And I know that that watery glow with its trio of denizens is a far better math homework buddy than a TV or an iPad. I haven't quantified it, but my sense is that she is happier and more relaxed working there at her desk and therefore likely to spend more focused time studying. (How many pediatricians' offices have you visited that include a fish tank? Even if they can't articulate it, there's a Blue Mind reason.) And as we've seen, Julia isn't switch-tasking back and forth between the tank and her algebra; this is a perfect neurological/environmental symbiosis. That's an important thing to keep in mind: Blue Mind isn't a matter of exclusivity but compatibility. When we swim we are doing many things at once: kicking our legs, pulling with our arms, timing our breathing, looking ahead. That's why a little fountain on our desk at work can perform its wonders: it's not distracting us, it's helping us focus. (If Julia had an open fire hydrant on her desk, we'd be in Red Mind territory!)

Water, Nature, and the Optimum Brain

There were profound reasons for his attachment to the sea: he loved it because as a hardworking artist he needed rest, needed to escape from the demanding complexity of phenomena and lie hidden on the bosom of the simple and tremendous . . .

— THOMAS MANN, *DEATH IN VENICE AND OTHER TALES*

"Brain training" is a big deal these days, with websites and apps offering extensive, research-backed programs that use exercises based on neuroscience to help you improve cognitive function, memory, visual attention, processing speed, flexibility, and executive function.[62] "Think faster, focus better, and remember more," they promise,[63] and advocates of brain training tell us, "You exercise your body to keep it fit and young; you need to do the same with your brain!" Whether digital circuit training (no pun intended) works or not, "use it or lose it" is as true for the brain as it is for the body: exercising your brain can keep existing connections strong and vital while building fresh connections as you learn new things. The brain is truly like the rest of your body in another way, too: while it gets "flabby" with too little stimulation, it also is stressed by too much activity; it needs different kinds of "exercise" to increase its different functions, and requires "rest and recovery" time to consolidate its gains.

Earlier we talked about the phenomenon of directed attention and cognitive fatigue. Such overstimulation causes the brain to get stuck in a kind of mental overdrive, continually revving at high speed and making us feel wired and tired at the same time. This wired and tired state describes a majority of college students. These (mostly) young adults are in an age group whose cognitive abilities are maturing and consolidating. However, they also are in the midst of intense and very stressful learning environments that require a *lot* of directed attention (and, of course, a lot of technology usage to boot)—attention that actually puts a greater demand on their maturing brains and creates more stress than is experienced by the over-twenty-five crowd. (And the digital avalanche is no longer confined to high school students; children of nearly all ages are increasingly occupied with cell phones and tablets more than they are with jump ropes, action figures, and soccer balls.)[64]

Several years ago researchers did a study of seventy-two undergraduate students living in dormitories at a large midwestern university. The dorm rooms were grouped by the views from their windows: trees and a lake, lawns and buildings, and brick walls and slate rooftops. Researchers visited the students in their rooms and administered standard cognitive tests, including the Symbol Digit Modalities test, which measures attention, visual scanning, and motor speed, and the Necker Cube Pattern Control test, which assesses the capacity to direct attention and inhibit competing stimuli. Students whose rooms overlooked trees and the lake not only performed better on the cognitive tests but also rated their "attentional functioning" as more effective than that of all of the other groups combined.[65]

University of Utah psychology professor David Strayer is an expert in the wired and tired world of adolescents: he has done extensive research on distracted driving and the effects of technology usage on the human brain. In 2012 Strayer helped conduct a study to see whether time in nature (and time away from media and technology) would improve higher-level brain function. Ruth Ann Atchley and Paul Atchley worked with him to observe fifty-six people who participated in four- to six-day Outward Bound hiking trips, during which they had no access to electronic technology while in the wilderness. Half of the people were given the Remote Associates Test (which evaluates creative thinking, insight, and problem-solving skills) the morning before they set out on their trips; the other half were tested on the morning of their fourth day in the wilderness. The people who took the test during the hike scored 50 percent higher on the test than those who took it before they left. "The current study is unique in that participants were exposed to nature over a sustained period and they were still in that natural setting during testing," the researchers wrote. "Despite the challenging testing environment, the current research indicates that there is a real, measurable cognitive advantage to be realized if we spend time truly immersed in a natural setting."[66] And it would seem that improved performance from spending time in natural surroundings is not just a short-term phenomenon: Researchers from the University of California, Irvine, discovered that people who had been on a backpacking vacation in the wilderness performed better on a proofreading test (used as a measure of cognitive performance) for several weeks afterward. In contrast, performance on the same test by two other groups—one of which took a nonwilderness vacation and the other continued their usual routine—declined over time.[67]

"Nature serves as a source that renews our attention, reinstating cognitive functioning with natural elements that invoke affective responses," write the authors of *The Economics of Biophilia*. Attentional fatigue, they added, "caus[es] stress to slow the heart rate and breathing while simultaneously arousing digestion to raise energy levels.... The combination induces lowered concentration and decreased effectiveness."[68] Certainly it seems that resting our overworked directed attention circuits and utilizing the "involuntary attention" circuitry that seems to engage when we are in natural settings can help soothe the hyperfocusing parts of the brain. At the same time, however, natural environments engage our senses in ways that our regular, urban, enclosed environments cannot. When all of our senses are engaged with the sights, sounds, smells, feels, and even tastes of nature, our minds are caught up in what Stephen Kaplan calls "soft fascination":[69] the effortless, involuntary occupation of the mind. "Nature, which is filled with intriguing stimuli, *modestly* grabs attention in a bottom-up fashion, allowing top-down directed-attention abilities a chance to replenish," writes Kaplan along with his fellow University of Michigan psychologists Marc Berman and John Jonides.[70] And by now the evidence is strong that there are *no* interior or urban environments that come close to having this sort of enhancing effect.[71]

With this sort of disparity in mind, Catherine Franssen

conducted a laboratory experiment with rats that investigated the effects of a natural environment on cognitive performance. Most lab rats are kept in cages made of clear plastic and mesh, with about 140 square inches of floor space per cage. The floor of the cage is usually covered with some kind of wood shavings; there's normally some kind of feeding and watering device. But Franssen and her team of researchers added natural elements—twigs, rocks, green leaves—to the cages of some of the rats, and plastic toys to the cages of others. Lo and behold, the rats whose cages contained the natural elements exhibited improved cognition, better problem solving, increased neuroplasticity, and more boldness and exploration of their environment.[72] Fascinatingly, these were lab rats that had been bred in cages and had never been in a "natural" environment. Were the improvements due simply to a more enriched environment, or did the natural elements added to the cages appeal to some deeper hardwiring? Like a great deal of psychological research, correlation is much more common than definitive proof of causation, but the correlation in this case tracks perfectly with studies of our own brains.

What's most interesting about time in nature, however, is what happens to the brain when it's *not* actively focusing or engaged. Think back to the last time you got your Blue Mind on: in the shower or bathtub, or sitting or walking by the water. You probably weren't thinking about anything in particular; you let your mind wander to wherever it wanted to go in your relaxed state. Perhaps you caught yourself daydreaming as you gazed out over the sparkling waves or ripples on the creek, and you reluctantly pulled yourself back to the present moment. For a long time scientists thought there was nothing going on in the brain when we allowed ourselves to daydream or "space out." But now we know that in those moments the brain's *default-mode network* is incredibly active. In other words, the brain at rest is not really at rest at all.

M. A. Greenstein is a powerhouse of a woman and educator, with short, dark hair and bright blue eyes. Her institute, GGI, specializes in advocacy and design and is dedicated to advancing access to brain health and knowledge. At Blue Mind 2 she spoke extensively about the default-mode network. "Drift is the freedom to wander in consciousness, and it's quite possibly one of the most important keys to the actual functioning of our nervous system," she said. (Note the terminology: though its Middle English origins related to herding, nearly every modern definition of "drift" is nautical.) "Drifting takes us into the default-mode network: the network that's active unless we are paying attention to something. In other words, it's basically 'online' until we call on other areas of attention. And the default-mode network devours huge amounts of glucose and a disproportionate amount of oxygen."

That last point seems odd: why would a network be so active when we're not paying attention? Scientists now theorize that the default-mode network allows the brain to consolidate experiences and thus prepare to react to environmental stimuli. This default-mode functioning has also been shown to constantly "chatter" with the hippocampus, the part of the brain that is key to neuroplastic development and helps create new memories and new learning. "We're starting to see a picture of ourselves as nervous systems that need to drift, that need to wander so that the brain can process at a very efficient level the amount of information that's coming in, translating it into neurochemistry and experience," Greenstein concluded.

All of this makes the notion of daydreaming more

complicated—in a good way, because it means that the default network is key to creativity and problem solving. Chen-Bo Zhong of the University of Toronto, Ap Dijksterhuis from Radboud University in Nijmegen, and Adam Galinsky from Northwestern University's Kellogg School of Management have noted that "conscious thought can subvert the search for creative solutions—novel connections or ideas often insinuate themselves into the conscious mind when attention is directed elsewhere."[73] Indeed, a 2012 study showed scientifically what most people know intuitively: letting the mind wander off the topic of a problem will lead to more creative solutions than focusing exclusively on the problem itself.[74] By processing chunks of data it has gathered, and beginning to make connections and to store new pieces of information away (our memories are distributed, not stored at a single node), the brain is beginning to fuse together information from different areas, forming new connections.

How many times have you had an insight, new idea, or solution to a problem pop into your head, seemingly from nowhere? That's the default-mode network kicking in, allowing your brain to make connections between different elements to create something entirely new. And being around water provides a sensory-rich environment with enough "soft fascination" to let our focused attention rest and the default-mode network to kick in. It's no coincidence that Archimedes was in the bathtub when he deduced a method for measuring the volume of an object with an irregular shape: the Archimedes Principle). *Eureka!* (Greek for "I have found it") indeed.

"Along the shore's edge are things you won't find anywhere else," says Michael Merzenich. "The feel of the water, the smell of the ocean, the birds you see there, the curiosities you see there, the boats on the surface—these are things that are special to that environment. And all of them are inherently personally calming, rewarding, and intriguing." The little changes in a natural environment—the sound of waves or waterfalls, the negative ions in the air, breezes blowing by—engage the brain's reticular formation, which calculates the amount of alertness needed to deal with the new stimulation. Most of these small changes produce just enough alertness to keep us curious and aware, but not so much that we can't relax our focus and let the default-mode network hum along quietly in the background. As naturalist Konrad Lorenz wrote, "A man can sit for hours before an aquarium and stare into it as into the flames of an open fire or the rushing waters of a torrent. All conscious thought is happily lost in this state of apparent vacancy, and yet, in these hours of idleness, one learns essential truths about the macrocosm and the microcosm."[75] Albert Einstein once commented, "Creativity is the residue of wasted time." But if time spent daydreaming helps us be creative, is it truly time wasted?

Of course, there are times where focus/directed attention merges with the open awareness of "soft fascination" and positive emotion to produce a state of effortless concentration and enjoyment. At these moments, we become lost in what we are doing: we are "in the flow."

Finding "Flow" in Water

It is the full involvement of flow . . . that makes for excellence in life.

—Mihaly Csikszentmihalyi

Almost every surfer I've ever known can tell you the precise instant when they got hooked on the sport, that moment when they first experienced the state of flow on the water. They had learned enough to stand on a board and feel the power of a wave propelling them toward shore; skill and exhilaration and an exceptionally beautiful environment came together to create a dopamine rush they would seek again and again. "Moments such as these provide flashes of intense living," writes Csikszentmihalyi;[76] they are moments when we lose track of time, nothing else seems to matter, and we feel we are truly alive and at our best. Writer and leader of the Flow Genome Project Steven Kotler beautifully described this feeling in his book *West of Jesus:*

> I paddled fast to my left, angling toward the next wave, stroked and stood and felt the board accelerate and pumped once and into my bottom turn, and then the world vanished. There was no self, no other. For an instant, I didn't know where I ended and the wave began.[77]

Surfers, whitewater rafters, kayakers, swimmers—pretty much anyone who participates in active water sports—along with rock climbers, tennis players, artists, musicians, and creative people of all types can experience flow when a few specific conditions are met. First, flow involves an activity you enjoy on some level; otherwise you would not put in the effort to achieve the second condition of flow, which is to have achieved a level of competence at which you no longer have to think about your performance and can simply enjoy the activity. The initial time you get into a kayak, for example, you have to learn how to get in without tipping over, how to use the paddle, how to steer, and so on; you have to invest a certain amount of directed attention to learn the mechanics of kayaking. Once you master the basics, however, you can start to enjoy the sport, and your focus now shifts to refining your skills, possibly adding more techniques to your repertoire, pursuing greater challenges or new places to kayak.

This is a crucial element of flow: you need to feel that you are being challenged when you undertake the activity. While some surfers will go out whatever the conditions, it's when the waves are just tough enough that they really feel in the zone. The challenge causes them to merge awareness with action and bring all of their skills to bear. This stretching of one's abilities combined with a pleasurable activity produces the fourth element of flow: the loss of a sense of the passage of time. When we're doing something we're good at and that we enjoy, and yet we're being stretched by the demands of this particular instance of the activity, we become completely focused on the present moment of doing *this* thing at *this* time in *this* way. We are so engrossed in our sport or project or art or activity that nothing else seems to matter, and we lose all track of time.

Given his interest in sound and the brain, Petr Janata is unsurprisingly an expert on flow in a very specific context: how the neural systems of perception, attention, memory, action, and emotion interact when we play or listen to music.[78] Janata points out that when we perform music, we are actively involved in an experience that can easily lead to flow; that is, we are usually doing something we have some experience with, and usually it is a pleasurable endeavor.[79] At the Blue Mind 2 conference, he posited that the medial prefrontal cortex (the part of the brain associated closely with emotions, self-image, creativity, and insight) is particularly activated by water. The comparisons between the brain on music and the brain on water poured out easily and steadily. Indeed, Janata remarked, "The same aesthetic emotions associated with music—for example, joy, sadness, tension, wonder, peacefulness, power, nostalgia, transcendence—all come up when people talk about being near water."

Recall that being challenged is a big part of what focuses the mind into a flow state. If each wave were exactly the same, the novelty that helps us reach Blue Mind won't be there; we'll habituate. Luckily, each wave *is* different (thank you, water). But not every task balances the raw and the cooked, so part of what we need to do when getting our Blue Mind on is to make sure that the routine doesn't become *too* routine. Recalls Ellen Langer,

> We did a study with symphony musicians, who, it turns out, are bored to death. They're playing the same pieces over and over again, and yet it's a high-status job that they can't easily walk away from. So we had groups of them perform. Some were told to replicate a previous performance they'd liked—that is, to play pretty mindlessly. Others were told to make their individual performance new in subtle ways—to play mindfully. Remember: This wasn't jazz, so the changes were very subtle indeed. But when we played recordings of the symphonies for people who knew nothing about the study, they overwhelmingly preferred the mindfully played pieces. So here we had a group performance where everybody was doing their own thing, and it was better.[80]

This Must Be the Place

While Csikszentmihalyi and others have shown that humans can access the state of flow in different places and with different activities, for many of us there seem to be particular locations and conditions where it is easier to perform at our best, and in particular to access the "flow" state of creativity. When we immerse ourselves fully in sensory-rich surroundings, paying mindful attention to where we are and what we are doing, even something as simple as a walk through a park or a float along a river can engage both our directed attention and default-mode networks, giving us the experience of mindful yet restful focus. A few years ago David Strayer took five neuroscientists, a guide, a photographer, and Matt Richtel, a reporter for the *New York Times*, on a seven-day trip on the San Juan River in southern Utah. For five days there would be no cell phone reception or Internet access; the travelers would be without technology of any kind. The scientists were interested in seeing what would happen to their cognition with the only stimulation available coming from the river, its environs, and each other.

Over those next five days the constant interruptions of modern life receded as the travelers paddled through rapids (intense focus), hiked trails along the canyon walls (physical stress), and floated along calm stretches of water where they could admire the natural beauty around them. And, as Richtel observed, ideas started to flow freely. The scientists agreed that "something" was happening cognitively that seemed to clear their heads and open them up to fruitful discussions on a wide range of topics.[81] Was it due to a lack of distractions?

Their brains finally getting the rest they desperately needed? Immersion in nature and time on the water? The nighttime darkness revealing a staggeringly dense assembly of stars? A change of scenery, or simply getting away from their regular routine? All of these factors may have contributed; but regardless of the reason, it is clear that getting away, and getting away in nature, can help us be better at what we do.

But a long river trip with a group of scientists is not in itself science, and the significance of the changes experienced was a matter of debate for the group. The more skeptical among them weren't convinced anything lasting—personally or scientifically—would come of the excursion. Yet by the time they emerged, Art Kramer, an ambitious, accomplished, and extremely driven University of Illinois neuroscientist, noted that "time was slowing down" (for the first time since he'd been fifteen years old) and commented that he wanted to look at whether clearer thoughts and other benefits to the brain came from being in nature, the physical exertion of paddling and hiking, or a mixture. Todd Braver, a psychology professor at Washington University in St. Louis, was making plans to study the brain at rest and in nature, using imaging technology. Paul Atchley, the professor at the University of Kansas who told us about the challenges of driving and texting at the same time, had further insights into the addictiveness of digital stimulation. Steven Yantis, chairman of the psychological and brain sciences department at Johns Hopkins and a brain imaging expert, pointed to a late-night conversation "beneath stars and circling bats" as the backdrop to thinking about new ways to understand his investigations of cognitive control during task switching. All of the scientists recommended a little downtime as treatment for a cluttered mind. As the river flowed, so did the ideas, noted Richtel.

Putting the dots together, it's clear that while an extended river trip is great, even a little time outdoors can do wonderful things. Sculptor David Eisenhour described it this way: "Being in nature quiets my mind, and out of that quietness is where the real art happens."[82] Ralph Waldo Emerson said of Henry David Thoreau, "The length of his walk uniformly made the length of his writing. If shut up in the house he did not write at all."[83] That's been true for countless numbers of people, and with good neurological reason. The list of intellectual and artistic breakthroughs sparked by a wander or a swim is long indeed. Sometimes to get somewhere, you have to go somewhere.

David Pu'u is part native Hawaiian and all world-class waterman. He's a federally certified first responder and rescue boat operator—in other words, he's the water-safety guy who will get you out of trouble in pretty much any kind of marine environment. But David's often the one who got you into trouble in the first place, because in addition to all the above he's an internationally renowned photographer and cinematographer known for getting the "big pictures" of surfers at Mavericks, sharks off the coast of Hawaii, and whitewater rafters on the American River. David's photos appear in such magazines as *Sports Illustrated, National Geographic,* and *Time*. He knows wave hydrology and meteorology; he can figure out what the weather's going to do and how the waves will break around a reef or shoal. And no wonder: he's been around the ocean all his life, and spends about 250 days a year in the water. He shoots in, on, underneath, and near water; and you'll find him using almost every kind of moving vehicle—from jet skis, helicopters, and airplanes to boats and surfboards—to get into perfect position for the shot that, he freely admits, his intuition guides him to take.

But David's creativity reaches far beyond his ability to capture the perfect image. His background includes competitive surfing, cycling, swimming, and auto racing, as well as literature and art. He's a provocative thinker who's not afraid to ask probing questions of anyone, whether it's an audience of the top people at NOAA and the Department of Defense or a group of handpicked invitees to the 2012 Sea-Space Initiative (a conference that brings together some of the most innovative thinkers who study "inner and outer space").

Ultimately, as David points out, art is not about beauty or cognition but about communication, connection, and impact. "Beauty is just finding and connecting to someone's spirit," he says. "You try to communicate not what you saw, but what went on inside in your soul. You use a visual thing to inform the soul and educate the spirit. And that's what creates community, that's what creates communication, that's how we connect. And that's the goal of anything in art."

David's extreme in his relationship with water, but one of the most encouraging findings of neuroaesthetics is that artistic representations can inspire brain activity similar to what it would be like if you encountered the real thing. This is especially important when it comes to water. While nothing can compare to sitting by a river or swimming in a pool or wading along the seashore, simply looking at pictures of such watery places causes our brain to shift into Blue Mind mode. This transitive potency is profound—just ask Michael J. Fox—and we'll turn to it, and other ways you can bring the blue into your dry life, next.

8

Only Connect

Our lives are connected by a thousand invisible threads...

—Herman Melville

When Harvard Medical School psychiatrist and empathy researcher Helen Riess spoke at our third Blue Mind Summit, on Block Island, she opened her talk with a video of birds over Midway Island, in the middle of the Pacific Ocean. The skies were filled with seagulls and albatrosses, their wings spread, their calls echoing over the soundtrack of ocean waves and Spanish guitar. Birds were walking and sitting on a green hill; there were close-ups of white and brown gulls together, their heads hovering over speckled eggs in nests; brown and fluffy chicks, taking food from their mothers' mouths...

Then the image changed: atop grayish sand you could just identify the skeletal head and desiccated body of an albatross, and in the middle of the body, just where the stomach would have been, was a pile of plastic debris—bottle tops, scraps of plastic bags, rectangular bits of unidentifiable trash. The scene, captured by photographer Chris Jordan and cinematographer Jan Vozenilek, shifted to a nest with a lone egg in it, the rest of the nest completely filled with plastic waste. Next you saw a bird lying gasping, twitching, on the ground; in the next picture a human hand reached into the body of that same bird and pulled out handfuls of bottle tops.[1]

It was very quiet in the room after the video ended. Then Riess said, "Because of this clip, I will never throw a bottle cap randomly away again. That is how empathy works. Empathy takes impressions, images, sounds, and sights, and then something happens in the human mind that translates all of that into a feeling we can't get away from. Empathy is what connects us to all living beings." Empathy, she explained, involves higher-order reasoning capabilities that let us imagine how others are feeling even if we don't feel the same way ourselves. Empathy and reading others' emotions, intentions, and physical states are thus critical to our ability to function successfully within social contexts. Perhaps that's why empathy is hardwired into our neurochemistry and neurobiology. Helen's empathy drove her compassion, which led to her rethinking of those maligned little plastic caps (which now always remind her of the ill-fated albatross on Midway Island). "Empathy by itself is like an electric pump through which no water circulates, and it will quickly overheat and burn," wrote Matthieu Ricard (whom we'll meet a bit later under very different circumstances).[2]

About twenty years ago, scientists in Italy discovered that when macaque monkeys implanted with electrodes in the ventral premotor cortex F5 area[3] watched researchers picking up and eating peanuts, the neurons in the monkeys' brains responsible for the same actions lit up. Because these specialized neurons essentially "mirrored" the actions of others, they were called mirror neurons. Immediately scientists wondered whether people possessed the same kind of neurons, and if so, where they would be located in the brain. Subsequent studies using fMRI imaging of humans showed that groups of neurons in the inferior frontal gyrus responded both to watching and doing the same action.[4]

Studying mirror neurons has opened up new lines of research into the ways humans relate to each other. "There's been an explosion of scientific research into shared neural circuits, or what's going on in our brains when things are happening in other people's brains," observed Riess. "In other words, the substrate of empathy."

The existence of mirror neurons demonstrates that we are wired to connect to the idea that the physical and emotional processes going on inside of *you* have their resonance inside of *me.* "Based on . . . sensory inputs, we can mirror not only the behavioral intentions of others, but also their emotional states," writes Dan Siegel. "This is the way we not only imitate others' behaviors but actually come to resonate with their feelings—the internal mental flow of their minds. We sense not only what action is coming next, but also the emotional energy that underlies the behavior."[5] We read others' emotions long before we're even conscious of it. In another fascinating experiment, psychologist Ulf Dimberg of Uppsala University in Sweden studied the ability of subjects to react to angry and happy faces flashed on a computer screen. Not only did people automatically frown slightly when they saw anger and turn up the corners of their mouths when a happy face appeared, but they did so 500 milliseconds after the picture appeared—faster than the conscious mind could register the image.[6]

Humans also demonstrate high degrees of physiological

synchronization: we yawn when someone else yawns, laugh when they laugh, and, in the case of babies, cry when they hear someone else crying. "These sensory, motor and emotional processes are playing an important role in our understanding of others," says Art Glenberg, a professor in the Psychology Department at Arizona State University and an expert in what is being called "embodied cognition."[7]

Unfortunately, empathy also has its downside: when we see someone in distress and say, "I feel your pain," we sometimes mean it literally. Riess described an experiment conducted by neuroscientist Tania Singer. Singer recruited sixteen couples, and had the female member of each couple spend some time in an fMRI scanner. She then gave each woman an electric shock to her right hand and mapped the areas in the brain that were activated by the pain. Next, Singer let the women know that their spouses had just received the same shocks, and mapped their reactions. Some of the same areas in the brain that had lit up when the women themselves were in pain activated when they knew their spouses were in pain. The circuits associated with feeling pain physically were relatively quiet, but the networks having to do with the *emotional* aspects of pain lit up significantly.[8] As Riess pointed out, this experiment showed that while the experience of empathy can lead to a greater sense of connection and altruism, it also can produce emotional distress and even caregiver burnout.

It may seem strange that any of this has anything to do with Blue Mind. But while we've been talking about mindfulness in terms of its ability to boost creativity and focus, to help us contextualize our personal lives and better decide what is and isn't urgent, we haven't really discussed some of the core elements of the mindful state. Yet this ability to recognize a greater whole is, in the end, the very essence of Blue Mind. In order to shed our stress and distraction, we need to recognize that our lives are part of a larger natural system. The word "holistic" essentially refers to the idea that pieces together are more than the sum of their parts. Our eminence thus comes from realizing that the question mark is more powerful than the exclamation point, that all of our decisions should be understood relative to the inordinate and mysterious hospitality that water allows us to tap into. But in order to reach Blue Mind we have to go beyond ourselves—we need to tap into each other.

Connecting with Nature

> *We need the sun, the moon, the stars, the rivers and the mountains and birds, the fish in the sea, to evoke a world of mystery, to evoke the sacred.*
>
> —Thomas Berry, *The Great Work*

When scientist (and agnostic) T. H. Huxley was asked to write the opening article for the very first edition of *Nature,* in 1869, he declared there could be "no more fitting preface" than a "rhapsody" by Johann Wolfgang von Goethe. "We are surrounded and embraced by her: powerless to separate ourselves from her, and powerless to penetrate beyond her," Goethe wrote, and Huxley concludes, "It may be, that long after the theories of the philosophers whose achievements are recorded in these pages, are obsolete, the vision of the poet will remain as a truthful and efficient symbol of the wonder and the mystery of Nature."[9]

In study after study, those who choose to spend time in nature speak about its ability to make us feel more connected to something outside of ourselves—something bigger, more transcendent, and universal. Some of my favorite recent studies include a 2011 survey of 452 students in Edmonton, Alberta, which showed that feeling connected to nature led to greater feelings of awe, vitality, purpose, and more positive emotions overall.[10] In another study, people who viewed nature scenes and imagined themselves fully immersed in nature were more concerned with prosocial goals and more willing to give to others.[11]

What is it about nature that inspires this feeling of connection?

First, the most frequently mentioned "transcendent" aspect of the natural world is its beauty—an increasingly exotic beauty, in our Red Mind world. "Even the person whose sole experience with nature consists of lying on a beach and watching the waves will not be surprised that those who visit the wilderness list aesthetics as one of their main objectives," writes Winifred Gallagher in *The Power of Place*.[12] Perhaps because our ancient ancestors saw beauty in the shapes and colors of the natural world, our response to nature's aesthetics is deep—and often poetic. And the experience goes well beyond the visual: we come across unfamiliar (read: novel) sounds, smells, flora, and tastes that we would not encounter back home. This is the way author and wilderness guide Sigurd F. Olson described one of his most memorable and beautiful moments in nature:

> A school of perch darted in and out of the rocks. They were green and gold and black, and I was fascinated by their

> beauty. Seagulls wheeled and cried above me. Waves crashed against the pier. I was alone in a wild and lovely place, part at last of the wind and the water, part of the dark forest through which I had come, and of all the wild sounds and colors and feelings of the place I had found. That day I entered into a life of indescribable beauty and delight. There I believe I heard the singing wilderness for the first time.[13]

Nature generously bestows a grandeur that puts us in our place. When he was a teenager, neuroscientist Dan Siegel would ride his bike to the beach, walk along the ocean edge, and think deep thoughts. "I'd watch the waves and be filled with wonder—about life, the tides, the sea," he recalled. "The force of the moon beckoning the water, raising it up toward the cliffs, then pulling it back down beyond the rocky pools, back out to sea... These tides, I thought, would continue their eternal cycle long after I was gone from this earth."[14] Trees, grass, water, sand—all are familiar to us, yet the size and scale of nature can make us catch our breath and marvel at its power. In its age, majesty, and complexity, nature dwarfs us—and yet we are drawn there because it puts our humanity into proper perspective. We encounter nature in a very physical sense when we walk, hike, climb, sail, paddle, swim, run, ski, or snowshoe through it; as hiker Adrian Juric says, these elemental forces "resist the sense of self we have worked so hard to establish" and cut us down to size.[15]

A 2007 study asked participants to describe a time when they saw a beautiful natural scene and to rate the level at which they felt ten different emotions. Words like awe, rapture, love, and contentment were ranked highest; people tended to agree with statements like, "I felt small or insignificant," "I felt the presence of something greater than myself," "I felt connected with the world around me," "I was unaware of my day-to-day concerns," and "I did not want the experience to end."[16] When participants in wilderness expeditions in the United States were surveyed in 1998, fully 80 percent said they had a greater spiritual connection with nature as a result of their trips.[17] We realize what I like to think of as a positive lack of control, as opposed to the lack of control we feel in our overstressed, overwhelmed lives. Our inability to have power over our inboxes and bank accounts and waistlines (not to mention the economy and international conflicts) simply makes us feel worse about ourselves. But in nature we realize there is something so immeasurable, so magnificent, that it exists both with us and without us. That recognition can transform our sense of responsibility and renovate our list of priorities.

Recent studies have focused on the different neural networks that we use when focusing on things outside ourselves (the extrinsic network) and when focusing on self-reflection and emotion (the intrinsic, or default network). The brain usually switches between the two, but cognitive neuroscience researcher Zoran Josipovic discovered that experienced meditators could keep both networks active at the same time while they meditated.[18] Doing so lowered the wall between self and environment, possibly with the effect of inspiring feelings of harmony with the world. That ability to simultaneously hold awareness of self and other is called nonduality, or oneness in both Eastern and Western philosophies. There's a sense of connection with everything, of no separation, of being part of something infinitely large and wonderful. Senses are sharpened; you see, hear, feel, taste, and smell more fully. Feelings of happiness, contentment, bliss, awe, and gratitude arise for no reason—some spiritual masters refer to this as "causeless joy." There's a sense of timelessness, or time seems to slow to a crawl. There's a sense of wanting or needing nothing else. Some would call it communion with the natural world; some would call it the experience of God. Perhaps most people wouldn't even know to put words of any sort to it.

Meditation can bring us to this state, as can prayer and other spiritual practices. But many of us feel moments or even hours of that sense of oneness and spirituality when we interact with nature, especially with water and the creatures we find there. "One cannot help but develop some form of attachment to the various social and natural landscapes that one encounters and moves through in one's lifetime, and frequently the feelings one forms in response to a particular place can be especially strong and overwhelming," state Laura Fredrickson and Dorothy Anderson.[19] We become attached to our particular "piece" of nature and treasure it for the experiences we have had there: it becomes our "sacred space." Your sacred space may be an inaccessible bit of wilderness reached only by foot or canoe; or it may be amidst the waters themselves, as you fished, sailed, or slipped in and felt the power of the water beneath or around you. But whenever or however you enter it, you feel connected to something greater than yourself.

Psychologist Abraham Maslow believed that because man's "higher and transcendent nature" is "part of his essence,"[20] occasionally we can access the mystical consciousness William James described. Maslow called these moments *peak-experiences*, and described them as "non-striving, non-self-centered, purposeless, self-validating, end-experiences and states of perfection and of goal attainment."[21] Psychologists studying these peak moments believe that they share certain characteristics: a complete focus of attention; an absence of fear; a perception that the world is good; a feeling of connection and even merging with the environment; feeling humbled by the experience and fortunate to have participated in it; a sense that time and space have altered and one is immersed in the present moment; a feeling that the experience is real, true, and valuable; flashes of insight and emotions not experienced in daily life; and a realization of the meaningfulness of the experience and the significance for one's future life.[22] When we access these states, we see ourselves not as separate but as "embedded" in our relationships with everything in the world; we are part of everything, and everything is part of us.[23]

Many times such peak experiences involve pushing yourself past perceived limitations. Catherine Franssen saw this with skydivers and rock climbers; Jaimal Yogis and other big-wave surfers describe moments in the ocean when "the wave demanded such hyper-focus... there wasn't even time to differentiate between one's body and the wave."[24] On the South Fork of the American River in California, a white-water rafter described the experience like this:

> The top of the mountain finally gives up at the end of the peninsula that creates the S turn I admire so much. The velocity of the water increases dramatically, the negative ions in the air from the rapids changes everyone's attitude. As I approach the thunder, my muscles throughout my entire body come to attention—as always, I go through the rocks 100 yards upstream I call the Goal Posts, knowing that if I can float my boat through them, I'll be OK in Troublemaker. Approaching the final turn... I tense as I grip my

> oars, I totally relax my mind and go for the flow—punch the hole and slip by the rock. And like magic, another peel off the layers of life, off the old onion, exposing fresh flesh and a new perspective on life.[25]

This sort of expansive awareness—"a new perspective on life"—is almost inevitably common in such circumstances that combine the natural world and water.[26] Indeed, as a spiritual element of the natural world, there seems to be something particular about water that permeates humanity's consciousness. When seeking to describe the experience of wholeness, limitlessness, and eternity, Freud drew on his correspondence with French writer (and student of Eastern religions) Romain Rolland and called it the "oceanic feeling."[27, 28] Many of our spiritual and religious traditions feature water. In the *Tao Te Ching* (written somewhere between the sixth and fourth century B.C.E.), Chinese philosopher Lao-Tzu wrote, "Of all the elements in the cosmological construct of the world, Fire, Water, Earth, Mineral and Nature, the Sage takes Water as his preceptor." The Buddha likened life to a river that is always flowing, changing from moment to moment. Water is integral to the creation myths of ancient civilizations from Egypt to Japan. "The spirit of God moved upon the face of the waters" (Genesis 1:2, King James Bible). "We [God] made from water every living thing," (The Quran, *sūrat l-anbiyāa* [The Prophets] 21:30). "Darkness was hidden by darkness in the beginning, / With no distinguishing sign, all this was water" (Rigveda 10:129:3). Hindus consider it sacred to bathe in the Ganga, "Mother Ganges"; Christian pilgrims flock to the river Jordan and Lourdes; Islamic pilgrims visit the Zamzam well in Mecca while performing the hajj. Humans ritually use water to cleanse themselves of metaphysical pollution ("The sea can wash away all evils," says Iphigenia, declaring her brother Orestes cleansed of the murder of his mother),[29] and as a means of consecrating the living (baptism with holy water) and the dead (bathing the body before burial). For many indigenous peoples around the world, water represents humanity's connection to all living things. Elizabeth Woody, a member of the Yakima Nation in Oregon, says, "Water is a sacrament in our religious practices and overarching medicine. It is the central symbol of our cycle of ceremonies. Along the 'Big River,' the Columbia, we wake with a drink of water, and close out the day with a sip and prayer.... Water equals all life."[30]

In 2010, Ian Foster of the University of Montana did a study of the spiritual connection felt by people on canoe trips through the Minnesota Boundary Waters Canoe Area Wilderness (BWCA), which consists of approximately 1.3 million acres with 1,175 lakes and hundreds of miles of streams. Much of the BWCA is accessible only by canoe, yet every year more than 250,000 people visit it to hike, canoe, kayak, fish, hunt, or camp. Foster conducted his research by canoeing to different campsites in the area and asking people to describe their experience of the wilderness. "Rather than standing at the trailhead after taking my morning shower and asking them about their trip and experiences, I was there, in a wild landscape," Foster wrote. "[I] had bathed in the lakes, caught fish for dinner (albeit twice in thirty days), paddled into the winds, and combated the same swarms of mosquitoes."[31] He discovered that it was in the beauty and quiet of "plateau-experiences" that people felt the closest to spirit. One man, "Tom," talked of soaking in "everything—the water, the trees, the sky the breeze. . . . I just turn off everything else and just soak in what is around me and take time to be thankful for it."[32] Being immersed in the natural experience, with limited social contact and cultural input, and required to interact with nature in much the same way that people native to the area had done for thousands of years—in such conditions, Foster commented, people's connection to something greater than themselves and to their surroundings was "kindled, stoked, and/or sustained."[33]

In descriptions of their spiritual connection to their environment, Foster discovered that water consistently played a significant part. The natural beauty of water and sky (in the Dakota language, *Minnesota* means "where the water reflects the heavens") touched many people. "Mary" described one such encounter:

> Yesterday we stayed at a campsite on Hudson Lake and the sky was this bright pink and purple, and it looked water colored—so amazing, like it couldn't even be real. . . . As the night gradually came on, the sky was getting darker and the water took that on, and I was just watching these two mediums entirely change all the time. . . . In that moment you are like, "Why am I here? What put me here in the spot so that I can feel this?"[34]

Peak and plateau experiences in nature are remarkable not just for their momentary impact, but, more important, for the effects they have when we return to our regular lives. In the middle of a busy day, on the streets of a large city, or in an office, with our eyes locked on the screen of our smartphone or tablet or laptop, taking a moment to remember a transcendent moment when the mind calmed and the heart opened to the beauty and wonder of nature can transport us back to the experience of feeling connected with nature, spirit, the divine, or whatever inadequate name we give it. "Nancy" summed up her own return experience: "I grasped something out there.... It's like everything is all right. This kind of deep sense of happiness, just by thinking back on it, is so powerful."[35]

The Million-Dollar Fish

Fish are friends, not food.

—Bruce the Shark, *Finding Nemo*

The Tsukiji Market in Tokyo is the world's biggest wholesale fish and seafood market. Every day (except Sundays and holidays) 1,600 vendors sell more than 400 different varieties of seafood, from seaweed to whale. The most prized fish of all is bluefin tuna. Approximately 80 percent of the bluefin caught anywhere in the world makes its way to consumers in Japan.[36] Bluefin tuna used to be ground up for cat food because no one wanted to eat it. But starting in the 1970s, it became a delicacy—especially the belly loin cut known as *toro*. According to seafood wholesaler Catalina Offshore Products, "The high fat content of bluefin *toro* results in meat that is pink to white color with a rich, buttery taste that melts in your mouth."[37] At the best Tokyo sushi bars, a single piece of top-grade *toro* can sell for around $24.[38]

I first visited the market in summertime, but at Tsukiji, it's considered an honor to buy the first bluefin tuna of the year, so the New Year's Day fish auction attracts a lot of interest from restaurateurs looking to gain status (and publicity) from a winning bid. In 2013, the winner was Kiyoshi Kimura, owner of a Tokyo-based restaurant chain, who paid $1.76 million for the first bluefin of the year. The tuna weighed approximately 489 pounds, so Mr. Kimura shelled out around $225 an ounce for his prize. However, he reported, the fish would probably bring in only about $4.60 per serving once it reached the customers in his restaurants later that day.[39]

What businessman would buy something when he knew he would never even come close to recouping his investment? Simple: the jaw-dropping sale price was mostly a matter of national pride, marketing, and good feelings, explained Kimura as he cut up the fish, holding aloft its torpedo-like, silver head to a sea of flashing cameras.

According to a report published just twenty-two days after Kimura's record-setting purchase, Pacific bluefin tuna numbers were down 96.4 percent from unfished levels.[40] A further dismaying statistic is that 90 percent of bluefin are caught before they reach an age where they can reproduce.[41] Paying that much for the first bluefin of 2013 was great publicity, free advertising, and a chance to intimidate the competition by demonstrating the size of your bankroll, but as the director of global tuna conservation at the Pew Environment Group, Amanda Nickson, speculated, "You have to wonder what that *last* fish is going to cost."[42]

Why are we willing to pay so much for bluefin tuna, shark-fin soup, or turtle eggs even though (or perhaps because) these creatures are rare or endangered? The million-dollar tuna is a clear example of how the search for immediate gratification, increased status, and our inability to envision long-term consequences can result in what I call the *neuroscience of destruction.* And is it possible, as David Pu'u says, that with greater self-understanding and self-awareness we can make better choices that can lead to a better future—choices that not only preserve our natural world, but the chance for more access to Blue Mind?

Our Psychological Relationship to the World

> *We want the maximum good per person; but what is good? To one person it is wilderness, to another it is ski lodges for thousands.*
>
> —Garrett Hardin, "The Tragedy of the Commons"

For about five years the city of San Diego has been running a Think Blue San Diego media campaign, including a series of commercials that ran on TV and in local movie theaters. One of the commercials, entitled "Karma," showed what would happen if someone who threw a piece of trash on the street was affected directly by everything that washed down to the ocean through storm drains. The businessman who has thrown away a gum wrapper while walking (1) gets doused by a soft drink thrown by a woman in a car, (2) steps in dog poop, and (3) is covered in kitchen garbage that another woman throws from her balcony.[43] "Think Blue, San Diego," the voice-over exhorts.[44] It's an amusing way to show the effects of a small action—effects that we usually don't think about because we can't see them directly.

The brain has an amazing ability to hide a world of truth from us. We're surrounded by billions of feelings, tactile senses, memories, sounds, smells, and a barrage of voices, and most of the time the brain insulates and protects us from much of it. But that thick padding comes with a cost: it means we really have no idea—most of the time—why and how we do what we do. Our brains excel at rationalization and self-deception, and these tendencies are hardwired into our cognitive apparatus.

According to a theory developed by the Swiss psychologist and philosopher Jean Piaget, when you hide your face from a baby, he or she thinks you're truly gone—the object (your face) is not a permanent part of the baby's perception. When you reveal your face again, the baby is surprised and (hopefully) delighted. At a certain point, however, the brain develops enough to realize that whether hidden or not, an object is still there. While there are disagreements about the age at which babies understand "object permanence," I believe a version of it affects our relationship with water and nature. In the "Think Blue" video, the man who threw out the gum wrapper, the woman who tossed the drink, and the lady who dumped the kitchen garbage without looking where it landed were all subject to the "out of sight, out of mind" principle. If we don't see our trash washing down storm drains to the beach, we don't worry about it. If we don't see homeless people on our way to work, most of us pay little attention to the homeless issue in our community. Unless people feel they and their families are threatened directly—by global warming, water pollution, beach erosion, or toxic waste, for example—then it's not likely they will pay much attention to the issue.

The flip side of "out of sight, out of mind" is also an example of cognitive blindness—if the environment where we live seems okay, then we don't really believe there's a problem. Barton Seaver is a National Geographic Fellow and director of the Healthy and Sustainable Food Program at the Center for Health and the Global Environment at the Harvard School of Public Health in Boston, and he points out that because we can go down to our local supermarket and see cod on sale every day, it's hard for us to believe that cod stocks have been decimated. If we visit a river or stream and the water looks and smells clean, we may not believe that the levels of pollution in our creeks, rivers, and streams are rising and the water is unsafe. No need for conservation when you've got plenty of water, right?

The most fundamental evolutionary force is the urge of every human to survive. For millions of years, when there were far fewer humans, we could exploit natural resources without the fear of exhausting them. We didn't have to think much about ecology or sustainability; we could just move on to somewhere else. We were concerned about survival of ourselves, our families, and our species, in that order. And now, because the threats to our species often feel distant and intractable, and survival seems theoretical rather than practical ("Temperature rise? Coastal flooding? A hole in the ozone?"), we find them hard to take seriously enough to do anything about them.

In his famous 1968 paper "The Tragedy of the Commons," the American ecologist Garrett Hardin stated that people will do things that are against the common good if (1) it will provide them with a greater benefit, and (2) they think they can get away with it. A lobsterman might set out more traps than he's allowed or put traps in protected areas, for example, because it will bring in more money for his family. We are "hardwired to be very self-centered and self-biased... good consumers but not good conservationists," comments Michael Soulé, one of the pioneers of conservation biology.[45]

The same year that Hardin wrote his paper, social psychologists John Darley and Bibb Latané performed a famous experiment: they sat students in a classroom and then slowly filled it with smoke that suggested that there was a fire nearby. If a student was alone in the classroom, he or she usually left quickly. However, if the student sat with a group of two to four other people (who were part of the experiment), and those people did not leave when the room filled with smoke, the student often remained—even if the smoke was so thick he or she could no longer see the other people in the group. Embarrassment and confusion were the two reasons cited by the students for not moving, but thcy also said they were looking to the others in the room to signal if there was anything wrong.[46]

The people we are with and the groups of which we are part are the strongest external factors influencing our decisions. Mark van Vugt theorizes that this is part of our hardwired response to the world: starting from birth, we unconsciously copy the behaviors of others around us. But not only the actions: we also imitate their beliefs, views, and decisions. Van Vugt points out, for example, that while most homeowners will say that the conservation behaviors of their neighbors has little effect on their own, the "greenness" of our neighbors is one of the strongest predictors of our own energy and water use. [47]

On the flip side, we tend to cognitively separate ourselves from those who are not part of our family, group, or tribe. "We all have a compassion blind spot—people we do not see as fully human; whose suffering is not as real as our own, or not as deserving of compassion as our own," says Stanford's Kelly McGonigal, describing to attendees of the first Blue

Mind Summit a 2007 study of how the brain reacts to people we consider part of extreme "out-groups." Instead of triggering emotions like pity or compassion, out-group individuals trigger activation in regions of the brain associated with disgust and threat consistent with aversive responses to rotting food.[48] It's part of our evolutionary cognitive programming to feel repugnance in such circumstances, but this tendency often prevents us from feeling either compassion or any kind of responsibility to help cure the problem.

Empathy, Compassion, and Oneness

Compassion is the keen awareness of the interdependence of all things.

— Thomas Merton

Matthieu Ricard is a large man with a shaved head, maroon-and-orange robes, and a warm, beautiful smile. A Buddhist monk as well as a photographer, philosopher, and author (who holds a Ph.D. in molecular genetics), he has been the subject of several studies of the neuroscience of happiness, meditation, and mindfulness, and a few years ago he agreed to participate in an experiment to examine the neural signature of compassion. A research team led by Tania Singer asked Ricard to immerse himself in what they describe as three different forms of compassion (nonspecific, compassion for the suffering, and a general loving-kindness) while his brain was scanned using fMRI imaging. The researchers were surprised to see that in response to images depicting human suffering, none of the "empathy for pain" circuits in Ricard's brain were being activated. Instead, his brain lit up areas associated with positive emotions, affiliation, love, and reward. Intrigued, the researchers asked Ricard to agree to be scanned again while he focused on emotionally sharing the pain of others' suffering. As he was being observed, Ricard brought up the memory of watching a documentary on children in a Romanian orphanage. "These children were completely emaciated and emotionally abandoned," he recalled. He promptly exhibited emotional empathy, his emotional pain network firing strongly. After a very short time, not surprisingly, he felt emotionally exhausted and "burnt out." Ricard was then given the choice either to leave the scanner or to continue the experiment while attempting to shift his feelings from pain to compassion. He continued.

While empathy and compassion are on a continuum of connection, compassion moves from "feel your pain" to "heal your pain." Indeed, as Ricard's scan progressed, his fMRI readings indicated that he had "turned off" the pain circuits and "turned on" the most positive areas of the brain associated with compassion. "Engaging in compassion meditation completely altered my mental landscape," he said. "Although the images of the suffering children were still as vivid as before, they no longer induced distress. Instead, I felt natural and boundless love for these children and the courage to approach and console them. In addition, the distance between the children and myself had completely disappeared."[49]

This desire to act links compassion to altruism, that is, actions to benefit others without the expectation of personal benefit. Just as we instinctively imitate others and sense their emotions, we also seem to have an instinctive desire to help others. A series of studies conducted by researchers from

Harvard University's Laboratory for Developmental Studies and the Max Planck Institute in Germany have shown that children as young as fourteen months spontaneously engage in helping behaviors, such as picking up a clothespin when an adult "accidentally" dropped it and then handing it back to the adult. These behaviors occur almost immediately, whether the parent is present or not, and even if the children must overcome obstacles or forsake playing with a new toy to help.[50] By now, you should expect that such actions involve neurochemical reinforcement, and that's certainly the case here: compassion and altruism have been linked to the release of beta-endorphins, the opioid hormones present when we experience feelings of love, warmth, caring, and social bonding. According to neuroscientist Joshua A. Grant, studies at the Max Planck Institute show that "brain regions involved in opioid signaling are active during compassion."[51] This may explain some of the broader benefits we receive from exercising compassion and altruism, which include better physical health and improved stress response.[52]

Matthieu Ricard is obviously an exception. But fortunately, you don't need to be a Buddhist monk with a molecular genetics Ph.D. to train your brain to change empathy into compassion. Even better, that training can happen in a very short period of time. A 2008 Stanford study found that as little as seven minutes of "loving-kindness" meditation increased feelings of closeness and social connection—a seven-point scale measured connectedness, similarity, and positivity—between subjects and photos of complete strangers.[53] On the neuroscience side, those who practiced loving-kindness meditation or participated in compassion training have demonstrated greater understanding of others' emotions and greater executive control.[54] Sounds a lot like Blue Mind.

The natural inclination of compassion is to extend and expand outward. As psychology professor David DeSteno commented, this is the goal of most meditation practices in spiritual and secular traditions: to "break free" from concepts that divide us and to view all creation with compassion and love.[55] When we do so, we can reshape our brains to tap into the experience of unified consciousness, or oneness with all things—including Blue Mind at its most powerful.

Hearts and Minds

No sensible decision can be made any longer without taking into account not only the world as it is, but the world as it will be.

— Isaac Asimov

"Neuroscience tries to understand two questions," says the director of Duke University's Center for Cognitive Neuroscience, Scott Huettel. "We try to understand why we value something—why there are some things we care about—and how we make trade-offs between the things we value." Value is assessed based on both internal and external factors. The emotional brain is hardwired to overvalue instant gratification and undervalue future rewards. Vanderbilt's David Zald, whom we met earlier when discussing addiction, tells us that studies in rodents show there are twice as many cells in the orbital frontal cortex that respond to immediate reward as cells that respond to the signal for a delayed reward. This is why we will eat the chocolate cake that's in front of us even though we have a longer-term goal of losing ten pounds, and why we find it hard to save for retirement when we're in our twenties and thirties. It's also one of the biggest barriers to making long-term changes that will decrease our carbon footprint.

So what will make us change our destructive ways? Fear can be a great short-term motivator, but the muffling power of habituation and the damaging distraction and suffocation that come from its Red Mind qualities makes us choose "flight" over trying to stay and do something. Jacques Cousteau once said, "People protect what they love." When you keep crying "Wolf!" about the environment again and again, when you tell people the world's going to end tomorrow and they wake up and it's still there, they stop paying attention.

People won't respond positively when we bore the heck out of them with a lot of facts. They won't respond with action when we make them feel guilty or bad about themselves. Yet that's exactly what the environmental movement has been doing: scaring people, making them feel bad, and overloading them with data.

Luca Penati is managing director of content and social at Ogilvy & Mather, and as an expert in shaping brands for companies, he understands what moves opinion and action. At Blue Mind 1 he said, "We know from painful experience that the most potent facts and rational arguments for conservation have mostly fallen on deaf ears. Storytelling helps us to communicate the urgency of the situation and the need for immediate action in a more powerful way. Storytelling is as old as history itself, and all great communicators have been brilliant storytellers. Now, neuroscience confirms that storytelling has unique power to change opinions and behavior." So for those hoping to raise awareness and commitment, we must put together simple, emotionally compelling messages in the form of a picture or a story. We don't want to fall asleep every night to nightmares of a dead ocean, burning rivers, and toxic tap water. As Barton Seaver comments, "People didn't come into my restaurant to be told 'bad, human, bad'—they came in for the joys of food. We want the human relationship with nature to be based on abundance, a relationship that encourages us to participate in the resilience, the restoration, and the reciprocal relationship that we need to have with nature."

Our (Layered) Relationship with Water and the World

> *The yearning to interact with the natural environment is inscribed in human nature.*
>
> —Eric Lambin, *An Ecology of Happiness*

As humans, we define ourselves in terms of our relationships—with family, social group, community, and nation. The Blue Marble photo reminds us of our most fundamental relationship: with our natural habitat. It's a relationship we may deny and ignore, but ultimately it is not something we can really take for granted. We must exist within this physical space, drawing upon its resources, interacting with it in ways big and small. We smile because the sun is shining, or feel depressed when there are too many days of rain or snow.[56] We marvel at nature's beauty in a flower, a raindrop, a waterfall, a mountain.

We watch in horror as tsunamis, hurricanes, floods, or tornadoes wipe out entire communities in minutes. We enjoy a stroll in a park or a bicycle ride through the country. Perhaps we sunbathe on a beach, or an apartment patio, or an urban rooftop. We visit aquariums or zoos, or watch nature documentaries. Unless we live in hermetically sealed environments far underground, away from any natural light, water, or air, we cannot escape our relationship with the natural world.

Our relationship with nature can be described as *egocentric, anthropocentric,* and *biocentric.* An egocentric orientation sees nature only from the perspective of what it can do for me personally. Can it provide food for *my* dinner plate, clean water for *my* shower, or a place for *me* to surf or boat or swim? Anthropogenic centers around the human side of the equation. It evaluates nature from the perspective of how humanity interacts with it—and how nature serves humanity's needs and desires. The anthropogenic view would look at the Gulf of Mexico in terms of the oil that can be extracted from beneath its seabed, the annual haul of shrimp, redfish, and oysters from its waters, and the value of tourism along its shores. On the other hand, a biocentric perspective sees humanity as part of nature, rather than separate from it. We interact with nature as with an equal partner, so to speak; we may look at the human need to harvest fish, for example, but we also take into account the effects of our fishing practices on the species we eat and the marine environment in which we cast our nets. We move from "humanity's needs first" to the recognition that we are participating in a dance of interdependence with our planet and its denizens, and that caring for our partners is, in fact, caring for ourselves. We understand that we have an interdependency that runs deeper than ecosystems, biodiversity, economics. Nature needs us, and we need nature. This enlightened self-interest is at the core of our very existence.

What creates this understanding? What switches us from egocentric or anthropocentric to biocentric? The same factor that moves any relationship from separate to intimate: love. I believe that this natural love is innate within humanity, bred into our DNA. As ecopsychologist Theodore Roszak describes it, we have an "empathic rapport with the natural world that is reborn in every child and which survives in the work of nature poets and landscape painters."[57] There also is an aesthetic element to our love of nature. We can marvel at the glories of the ocean's might (is there a surge of feelings quite as divine as those that come from standing on a cliff just above the waves crashing upon the Cornwall coast?), majestic trees, broad rivers, clear streams; we can see and appreciate the incredible variety of plant and animal life with which we share the world. When we are exposed to nature, we are programmed to be fascinated by it, and often that fascination turns into an emotional bond.

"When we walk the beach, play in the surf, or go for a swim in our favorite slice of the sea, we get in touch with what's going on there. We begin to care about that place. We become invested," writes river activist Christopher Swain.[58] With immersion in nature, our attachment deepens. Immersion moves us from disinterested appreciation to active participation. "The more experience a person has with nature, the stronger the pull toward it," writes Stephen Kaplan. "I argue that our species has a general preference for and valuing of natural stimuli, because with a very few units of experience, a great many people find a natural environment to be profoundly affecting. When people who've had little previous outdoor experience say they want to make it an important part of their lives in the future, as often happens following participation in a nature program, that's compelling. In a short time, something important happens to all kinds of people in natural places."[59]

Nature is "not just a heap of disjunct parts, but dynamic, intricately organized and balanced systems, interrelated and interdependent in every movement, function, and exchange of energy," writes ecologist Joanna Macy. "Each element is part of a vaster pattern, a pattern that connects and evolves by discernible principles."[60] When we seek to understand more about the natural world around us, when we can begin to see the interconnections between diverse kinds of life, between ocean currents and shorelines, between us and all the creatures and climates that exist on our blue planet, a deeper relationship is born. And from that deeper relationship arises an instinctual desire to protect what we love.

Outside-In: We Are Affected by What We Experience

To heal the ocean, we must heal ourselves.

— Dr. Rod Fujita

On the Fourth of July 2010, I flew into New Orleans and watched from above as fireworks lit the night sky—I'd never seen fireworks from that angle before. A few hours later, I was flying over something else that I'd never seen before but hope never to see again. Our small Cessna traced the coastline of Louisiana and Mississippi, documenting the flow of oil and tar balls onto islands, wetlands, mangroves, and beaches, and the inadequacy of the bright yellow and orange booms floating here and there but more often than not beached, twisted by the wind and waves. This industrial disaster was the result of the Deepwater Horizon oil spill, the largest in U.S. history.

We turned offshore, looking for deep water. Beneath us lay muddy, oily water, and then we could see the edge between deep blue and shallower oily water extending into the distance. I've spent my adult life working for the ocean, with the endangered animals living in it, and beside the people who depend on it. I've seen the wholesale destruction of species by commercial fishing, illegal hunting, and the destruction caused by plastic pollution. But none of that prepared me for the sight of thousands of square miles of destroyed ocean habitat—and the smell of bands of oil extending off into the distance, demarcated by the deep blue of the Gulf. I still think of Deepwater Horizon every time I pump gas into my tank or ride my bike behind a truck on a busy street.

I spent two weeks helping with the rescue and rehabilitation of sea turtles, pitching in however I could. During the BP oil spill and immediately afterward, 4,080 birds and 525 sea turtles were found dead. More than 16,000 miles of coastline from Texas to Florida were affected. And the human toll was equally bad—the physical, psychological, ecological, and economic impact was massive. Thousands of fishermen, rig workers, and people in the tourism industry lost their jobs. More than 174,000 businesses and individuals filed claims with the Gulf Coast Claims Facility for losses due to the spill.[61] People complained of coughs, nosebleeds, itchy eyes, sneezing—all symptoms of crude oil exposure, according to the director of the environmental and occupational health sciences program at Louisiana State University. Others who were involved with the massive cleanup operation reported chest pain, dizziness, respiratory problems, and gastrointestinal upset, which may be due to exposure to the chemical dispersants used to break up the oil. Psychologically, people fared no better. Mental health professionals in Louisiana reported increases in post-traumatic stress, anxiety, and depression; upticks in calls to mental health and domestic violence hotlines; and admissions to women's shelters.[62] Stefan E. Schulenberg, professor of psychology and director of the Clinical-Disaster Research Center at the University of Mississippi, who has made an extensive study of the psychological effects of disasters like the Gulf oil spill and Hurricane Katrina, observed symptoms in clinic patients including anger, irritability, stress, anxiety, depression, disruptive eating and sleeping patterns, and drug and alcohol use.[63]

Researchers examining the effects of the spill on directly and indirectly affected communities in Florida and Alabama found that while residents in all communities displayed clinically significant depression and anxiety, the highest levels of psychological stress were linked to loss of livelihood and income.[64] Several chapters back, we talked about the love most fishers have for their occupation—a love that keeps them working in the most dangerous occupation in the country. The people who live along a coastline often come to love the beaches, the breeze off the water, the endless horizon, the constant variety of wave, surf, and sand. Now, imagine a place that you love fouled and smelling of oil and covered with black goo. Think of how helpless and angry it might make you feel. Imagine that the very thing that provided all the watery benefits we've talked about throughout this book now produces feelings of disgust and may make you ill. I know that the Gulf oil platform workers, oystermen, barge captains, helicopter pilots, sport fishing guides, and beachcombers all felt the same deep pain and profound loss that I did, only much, much more.

Or perhaps, like some people, you take action: around 55,000 workers and volunteers participated in cleaning up immediately after the Gulf spill. The people I saw participating in that cleanup may have been sad, or angry, or determined, or just plain "business as usual" blasé—but when we talked among ourselves after spending the day up to our elbows in black gunk, shoveling sand, or cleaning sea creatures and birds in the hope they could be restored to health, you could tell that everyone involved had a deep connection with nature and water that lit them up inside. We would swap stories of our favorite beaches and the most magical moments we'd ever spent in, on, or around the water. This, too, is a sort of communion.

"Today we participate almost exclusively with other humans and with our own human-made technologies," writes David Abram. "We still need that which is other than ourselves and our own creations.... Direct sensuous reality, in all its more-than-human mystery, remains the sole solid touchstone for an experiential world now inundated with electronically-generated vistas and engineered pleasures."[65]

Without connection, there is no emotional drive to care for the world. Without connection, the egocentric and anthropocentric views take over, and value in the natural world is defined only by what can be put to use by humans. What's worse, considerations of the emotional, cognitive, spiritual, creative, and health benefits of water and nature are left off the balance sheet entirely. These elements make us better human beings, but because they are not measurable, they're not valued.

Ocean advocate Jean-Michel Cousteau, son of Jacques-Yves, always says, "When we protect our waters, we protect ourselves." Human health is intimately linked to environmental health. If our food, air, and water aren't well, neither are we. But there's much, much more to it than that. The vast benefits that people around the world derive from healthy waterways go beyond the waters we drink and the fish we eat. What we receive from healthy water is invisible, personal, intimate, and something I believe we can't thrive without. My hope is that as we fill in the blanks with research and further conversations, our understanding of how good it is for us to be by the water will become a powerful argument for keeping our world's waters clean, healthy, and free. But understanding is never enough. Humans make decisions and take action based on emotion, first and foremost. So when it comes to making decisions to care for our world, we must engender powerful emotions, and the best way to do that is by the stories we tell.

Telling the Story of Water

Don't talk to their minds; talk to their hearts.

— Nelson Mandela

Granted, a few people respond to the combination of fear, guilt, and doomsday information; but those who are left have no desire to be part of the club. They see many "doom and gloom" environmental activists coming and think, "You make me feel sad and bad about myself. You make me feel guilty for brushing my teeth and taking showers that last more than five minutes, or eating the wrong kind of fish, or washing my car in the driveway because the runoff will pollute rivers and streams." We can't continue to inundate people with negative information as we try to change their minds about things like "sustainability," "overfishing," and "climate change." Social science research has demonstrated that new information can cause people to be even more entrenched in their beliefs.[66] Our intentions are pure, certainly; our hearts are in the right place. As David Pu'u observes, "So many times as communicators we see something incredible, we think we have the answer, we see a problem that we can fix, we want to rush out and tell everybody." The problem is that the problem-based approach isn't working. Assaulting people with new information doesn't work either. We need to do something different.

So, how do we tell the story of water?

In *A Sand County Almanac*, Aldo Leopold wrote, "We can be ethical only in relation to something we can see, feel and understand, love or otherwise have faith in." I believe in telling a different story, a Blue Mind story, about our relationship to the world. It's time we recognize what economists, marketers, and politicians have known for years: that deep-seated, inscrutable emotions, not simply rational thoughts, rule human behavior. Our deepest, most primordial emotions drive virtually every decision we make, from what we buy to the candidates we elect. We need to tell a story that helps people explore and understand the profound and ancient emotional and sensual connections that lead to a deeper relationship with water. The Blue Mind story seeks to reconnect people to nature in ways that make them feel good, and shows them how water can help them become better versions of themselves.

But we have to tell the right stories in the right ways. A teacher in Oakland, California, demonstrated an effective way to tell the story of our natural world. For several months, first graders at the Park Day School turned their classroom into an ocean "habitat." They made fish out of aluminum pie plates and colored paper, and papier-mâché octopi with dangling tentacles of brown paper. The walls were draped with blue tarpaulins, mimicking water; patches of red paper "coral" dotted the floor of the classroom. Students each chose different creatures to study and then presented their projects to the class. All the other children in the K–8 school came to tour the classroom; there was a special presentation in the evening for the class's parents. The kids loved it; everyone agreed that the project was a huge success. The next morning the children walked in to see their beautiful ocean habitat covered with "oil." Black plastic garbage bags draped the coral, the fish, the octopi, the art projects. "There's been an oil spill," their teacher told them. After a classroom discussion of oil spills and environmental harm, the children put on plastic gloves and cleaned up the "oil." It was a visceral learning experience for the kids, about the problems we face and also about the ways in which humans can help.[67]

The Big Splash

Throughout this book you've seen the many ways that water is good for us—physically, cognitively, sensuously, financially, creatively, productively, spiritually, and healthfully—even when our brains sense it only indirectly through art, sound, certain words, the flow of a pencil, or the texture of an object. But for me, and for millions of people everywhere, the best and biggest benefits of water are all emotional (provided we are sufficiently internally hydrated). We love being in, on, under, around, or near it. We plan our vacations to spend time with water, and dream of the next chance we can get to jump on a surfboard or into a boat, visit a beach or lakeside or urban pool, or even slip into a bath or hot tub at the end of a long day. Try as we might, no amount of scientific data, fMRI scans, EEG readings, or carefully designed research projects can really show us exactly what we feel at those moments. And while real estate agents and economists can calculate the water premium to the penny and ecologists can track the movements of water molecules on their journeys through the web of life, they can't put a price on the moments when we experience water's beauty firsthand.

"If you want people to care about water, take them to the beach," Michael Merzenich says. "And don't do it just once: do it twenty or thirty times at least across their childhood. Because each time you do that, they're going to incorporate their positive feelings about it into the person that they are." A few years ago, psychologists at the University of Toronto conducted a revealing experiment: they split volunteers into two groups and, replicating the conditions of a call center, gave each group instructions on how to ask for donations from those they reached. Both groups had the same instruction sheet except for one thing: one group's sheet included a photo of a runner winning a race. "Much to my astonishment, the group that saw the photo raised significantly more money than the control group," one of the psychologists said. The surprise was so great that the experiment was rerun repeatedly, always with the same results. But here's what was so amazing about what happened: when the members of the groups that raised the most money were asked about what effect the picture of the victorious woman had on their soliciting, the same answer came back: "What picture?"[68]

Water primes us to feel certain things without letting us know it inspired the relevant neurochemical reactions. Our joy about water can do the same when it comes to helping others recognize why protecting it is so important. If the picture in our brains is of the summers we spent by the pool, the dolphins we saw playing in the water, the lazy afternoons in a boat, the thrill of standing next to a thundering waterfall in the middle of a forest, frog-kicking our way toward an iridescent fish hanging out near a coral reef—our "solicitation" will be more successful even without our knowing why.

Such personal appeals are crucial to combating the abstraction that makes standard appeals unsuccessful. In his book *Give and Take*, Wharton professor Adam Grant writes about an Israeli study that showed that attaching a patient's photo to his or her CT exam increased the radiologist's diagnostic accuracy by 46 percent. Simply putting a face to the image caused the radiologists to feel a stronger connection to the patient.[69] The challenge is that water *isn't* a person, and often showing photos of polluted beaches or dead birds doesn't really create the kind of relationship a photo on a CT scan would. "People will form a stronger attachment to and will do more for an individual than they will for a group, so maybe we should be thinking of the ocean as an individual," suggested Petr Janata at Blue Mind 2. Indeed, when we tell stories of our own interactions with water we turn "water" into a personal, individual experience, and an unforgettable memory.

Here, too, Blue Mind can help. In 2013 researchers in the United Kingdom studied 104 visitors to the National Marine Aquarium in Plymouth to see what effects such visits would have on their attitudes and intentions. Half the people received informational booklets on behavioral solutions for overfishing, and the other half simply visited the aquarium. Based on before-and-after surveys, time spent at the aquarium caused improvements in all participants' attitudes toward marine sustainability; adding the booklet also improved intentions as well. "A visit to an aquarium can help individuals develop what we term a *marine mindset,* a state of readiness to address marine sustainability issues," study authors wrote.[70] The combination of direct experience and emotional engagement found in the aquarium was an essential first step; giving people suggestions for potential behaviors was an invaluable second step. A fish tank may not be equivalent to feeling the undertow suction sand from around your toes, but it's still a catalyst.

The stories we need to tell must be couched in plain language, without knots and tangles of scientific jargon. People are not stupid; they will understand even the most complex ideas as long as they are presented in direct ways and real language. Even when we are speaking of the intricate interactions of oceans and rivers and runoff and, yes, the neuroscience of why we love water, we need to make these concepts clear and show how they are applicable to all of us.

Psychologist Daryl Bem[71] theorizes that our perception of ourselves is shaped by the ways we actually behave, rather than merely our perception of ourselves. Every time we decide to put our trash into the bin instead of throwing it on the street, for example, it's yet another confirmation of our identity as someone who cares about the environment. We build our "self" through our actions.[72] This makes sense, because we know that consistent actions reshape the pathways in our brains. By offering stories that include a call to action, we help them build an environmental self-identity that influences our values and choices. Every small successful action, recognized and celebrated, makes us feel good and causes us to want more of the same. Such consistent actions can subsequently become habits. Then we no longer have to think about whether we bypass the plastic disposable water bottle in favor of the stainless-steel one to carry with us, or choose the sustainable seafood option on the menu, or carpool or take our bicycles to work as often as possible. In San Francisco there is a wonderful Vietnamese restaurant called the Slanted Door, owned by Charles Phan, who emigrated with his family from Vietnam in 1977. Charles owns seven restaurants in the Bay Area and is one of the top restaurateurs in the state. A while ago he decided to make his restaurants supersustainable. He cut plastic use, chose organic ingredients, started composting, and bought only sustainable seafood. When I asked him, "How did you get so far ahead as far as environmental issues are concerned?" he answered, "I'm not an expert on sustainability, but I talk to people who are. I get their best advice and apply it. By the way, what do you think could make my restaurant better?" Just then, a waiter walked by with a tray of drinks, each of them with a plastic straw. I pointed at him and said to Charles, "I'd make straws available only upon request." He said, "We'll start that on Monday!"

The Blue Mind choice can be that simple.

A Pocket Full of Ocean

I live in a place where blue and happiness intersect all year round. We call it the Slow Coast. It's about fifty miles of cool, peace, and quiet—just a jump south of San Francisco, isolated by a mountain range but just fifteen miles as the crow flies west of Silicon Valley and a few miles north of Santa Cruz—with a smattering of great organic farms, tons of wildlife, country lanes, comfortable lodging, forest trails, pocket beaches, state parks, and a scenic, rocky coastline. Summer temperatures usually hover around 70 degrees, traffic is minimal, and nothing and no one will hurry you along. If your idea of a good time is an unhurried walk on a windy beach, picking your own food and then eating it, chilling out with a glass of wine, listening to some live music, then hitting the sack early to wake for a wildlife hike or bicycle ride, the Slow Coast could be a place for you, too, to get your Blue Mind on.

However, it was on the Slow Coast that not long ago I got a glimpse into the other "blue mind"—the kind of "blues" that no form of water or watery happiness can erase. I had a phone call scheduled with a friend, Drew Landry, who works on the Gulf of Mexico coast to raise awareness through his music and advocacy about the lingering effects of the 2010 oil spill. I knew it would be an intense conversation, so I went out to my other "office," on the bluffs over the Pacific Ocean a short ride from my house. During the call I watched gray whales pass by on their way to Mexico, otters diving in the kelp beds, seals and sea lions, countless seabirds, and a family down on the distant beach with their dog, taking it all in. It was a glorious day: it helped me keep my equilibrium as Drew and I talked about the incalculable stress that has wrecked lives and families along the Gulf Coast and led to suicides in communities where for years water had been only a source of happiness, identity, and prosperity.

After the call ended, as I walked back along the trail that parallels the railroad tracks and leads to Highway 1, I passed a man carrying a box. He also appeared to be enjoying the coast, so I simply nodded and said hi. He nodded back, and I continued on my way to the Davenport Roadhouse, a restaurant on the other side of the highway, for lunch and another meeting.

During lunch I noticed the chef, owner, and manager of the restaurant all leaving in a hurry and heading down the path to the bluffs. I commented to my associate that they must be going to check out the whales. Several minutes later, however, the chef, Erik, came back and said there was a man lying on the ground on the other side of the railroad tracks. (The housekeeper at the inn on the second-floor balcony had looked out and seen him fall.) I got up and immediately headed back to the bluffs, thinking that perhaps my EMT training could be of use until emergency services arrived.

It was too late for that. The man was slouched over the tracks, face down, a .45-caliber pistol by his foot. He had shot himself in the chest, spun around, and collapsed.

It was the same fellow I had passed on the trail. The box he carried was a gun case.

As we stayed by his body until the authorities came, I wrestled with my feeling of responsibility for not helping this stranger. All I could think was that I should have invited him to go for a swim or a walk out to the edge of the bluffs. We could have talked about how beautiful the ocean is. I could have told him stories about the life history of the whales and the otters.

When the deputy arrived, he told me that it's not all that uncommon for people to come to the coast to commit suicide. It's difficult to get up the nerve to kill yourself, he said, and ironically, the calm offered by the ocean helps people relax enough to do so.

My worldview was scrambled. The ocean *doesn't help everyone?* When you've loved water as much and for as long as I have, it's hard to wrap your head around the fact that other people don't feel the same. Amir Vokshoor, the medical researcher we met earlier when discussing the positive effects of the color blue, reminds us that while many people experience positive emotions around the ocean and other natural spaces, sociocultural influences play a part in how we respond to such stimuli. "Some individuals have a sense of awe about the ocean, while others, due to various experiences in their upbringing, have a strong fear response," he remarks. Stephen Kaplan adds, "It's terribly important to realize that within all people there's a very strong pull toward nature and also a fear of it.... People don't like to feel helpless or scared, and nature can certainly make us feel that way, especially if we're not equipped to deal with it."[73] Author David Foster Wallace described it as the "marrow-level dread of the oceanic I've always felt, the intuition of the sea as primordial *nada*, bottomless, depths inhabited by cackling tooth-studded things rising toward you at the rate a feather falls."[74] Even sound artist Halsey Burgund, who's spent years recording people's responses to being by water, has heard a few comments like, "You feel a sense of fear because there's just a vast expanse of water around you. It was mysterious because it went so far."

In his descriptions of what he calls "nature deficit disorder," Richard Louv points out that because so many children and adults spend their days in completely man-made environments, they develop aversions and even fears about what they view as dirty, untamed, and dangerous natural environments. More specifically, people who suffer from aquaphobia (irrational fear of water) may have had bad water experiences as children, or simply were terrified when faced with a completely unfamiliar element. "Every swimming teacher is eager to claim that water and swimming are great fun. I think they are lying," said Ilkka Keskinen of the Department of Sports Sciences, Jyväskylä University, Finland. "The water gets into your ears and your hearing is disturbed. It also tastes bad and . . . it feels very uncomfortable when it goes up your nose. And when it goes into your mouth and you breathe it, you probably think that you are going to die."[75] (As befits a Scandinavian researcher, Keskinen also fixates on frigid water temperature and how it "turns you blue.") Chlorine might kill germs, but it doesn't make our eyes feel better; salt isn't much comfort, either.

Yet an increasing amount of evidence is showing that even if we fear the untamed quality of nature, of wild waters, there is still something in us that needs natural connection in order to be happy. Louv tells the story of how his father, a chemical engineer who had once reveled in growing enough Swiss chard in his garden to feed the entire neighborhood and described that summer as "one brief Eden," became depressed and no longer spent much time outside. He retired to the Ozarks to garden and fish, but did neither. Eventually he, too, took his own life. "Which came first, the illness or the withdrawal from nature?" writes Louv. "I honestly don't know the answer to that question. But I often wonder what my father's life would have been like if the vernacular of mental health therapy had extended...into the realm of nature therapy.... As I watched my father withdraw, I wished that he would quit his job as an engineer and become a forest ranger. Somehow I believed that if he were to do that, then he would be all right. I realize now, of course, that nature alone would not have cured him, but I have no doubt it would have helped."[76]

I, too, share Louv's belief in the power of nature to help heal us and make us happier. Yet as I walked homeward after the coroner had taken the man's body away, I wondered whether water is a mirror for our darker emotions as much as it is an engine for our happiness. Water quiets all the noise, all the distractions, and connects you to your own thoughts. For many people, this brings feelings of happiness and calm. In the case of the man on the railroad tracks, perhaps the calm had turned into a final resolve. I didn't know a single thing about him other than that he was wearing a blue T-shirt and blue jeans, and he had shot himself on the Slow Coast. But I felt a need to let the man's family know that he seemed to be enjoying the beautiful day and that his very last moments were filled with sun and ocean and whales—and a quiet nod from a stranger.

A few days afterward, I stopped by the restaurant, and Erik said that the man's family had come by to retrieve his car. They mentioned that he had a terminal illness and had decided to end his life rather than suffer. In a strange way, that made me feel better. The man in the blue T-shirt had made an impossible decision, and he had come to the edge of the ocean to clear his mind, relax, and say goodbye on the most beautiful day of the year. I hoped that in his last moments there was perhaps a moment of, if not happiness, at least peace.

In the waves and the rivers and the lakes and the ponds we see what was, what is, and what is beyond us. Now we must figure out ways to make sure we are also seeing what will be.

Preserving, protecting, and restoring our waters are tasks for many lifetimes, and sometimes the effort can seem overwhelming. But as long as we stay connected with all of the many, many blessings that water provides, and continue to keep that love in the forefront of our minds and hearts, as long as we remind ourselves to hope, then our stories will help connect others to water and encourage them to do what they can to help care for this beautiful Blue Marble world. Blue Mind helps us strip away the anxious complexity and distraction of a Red Mind world. "If you just sit and observe, you will see how restless your mind is," Steve Jobs explained. "If you try to calm it, it only makes it worse, but over time it does calm, and when it does, there's room to hear more subtle things—that's when your intuition starts to blossom and you start to see things more clearly and be in the present more. Your mind just slows down, and you see a tremendous expanse in the moment. You see so much more than you could see before."[77] And there is no wider panorama than what we see when we look outward from the beach.

9

A Million Blue Marbles

The real voyage of discovery consists not so much in seeking new territory, but possibly in having new sets of eyes.

— Marcel Proust

As an ocean advocate and educator, I have been telling stories about water all my life, but I'm proudest of one particular story that has been passed from hand to hand around the world. It begins with a small, round piece of blue glass...

In 2009, almost thirty-seven years after the "Blue Marble" photo taken by the Apollo 17 crew, I stood outside the Simons IMAX Theatre at Boston's New England Aquarium, handing blue glass marbles to each person coming in to hear my talk.

People asked, "What's this for?"

"Hang on to it," I told them. "You'll find out."

In truth, I wasn't sure what I was going to do. I'd handed a blue marble to a friend, LeBaron,[1] at a cafe in Santa Cruz the week before and asked her how it made her feel. "Good," she said. "I like it." The image of the blue marble had been percolating through my brain ever since.

As I went through my presentation, the idea humming

along in the background (and I'm sure the blue of the marble helping my creative impulse), things got clearer. Then I came to the final slide in my deck—that famous Blue Marble photo—and I knew.

Pull out the marble you received when you came in, I told the audience. Hold it out at arm's length in front of you, and look at it. That's what Earth looks like from a million miles away: a small, blue, fragile, watery dot. Bring the marble close to your eye, and look at the light through it. Suddenly it's as if you're beneath the water. If that marble actually were made of seawater, it would contain trace amounts of virtually every element. It would hold hundreds of millions of organisms—plankton, larvae, single-celled creatures—in that one spoonful.

Now, I said, think of someone you're grateful for. Perhaps someone who loves the water, or is helping keep the planet's waters clean and safe and healthy. Or just someone you are grateful to have in your life. When was the last time you told them that you appreciate them, if ever? Think of how good it would feel to you and to them if you randomly gave them this marble as a way of saying thank you. It's such a simple thing that we're taught by our parents, to say please and thank you, but we don't do it often enough.

Take this marble with you, I continued, and, when you get the chance, give it to that person you thought of. Tell them the story of what this marble represents—both our blue planet and your gratitude. Ask them to pass the marble along to someone else. It's a reminder to us all to be grateful, for each other and for our beautiful world.

Will you do that? I asked.

They did. Over the next few days, I received a number of e-mails and comments from audience members, and I realized those weren't the last blue marbles I'd be sharing. I started doing the same exercise at other talks and presentations. I shared blue marbles with kids and adults. With no real budget but with the help of friends, I set up an organization and a website where people could get their own blue marbles for sharing. The number of blue marbles passing from person to person around the world—and the stories—grew exponentially, reminding people of what they're thankful for and encouraging them to say thanks. The Blue Marbles Project went viral: within eighteen months nearly a million people had received or given marbles to express gratitude for work that benefits our planet. James Cameron took a blue marble on his dive to the deepest part of the ocean.[2] Blue marbles have made it into the hands of Jane Goodall, Harrison Ford, Edward O. Wilson, the Dalai Lama, Jean-Michel Cousteau, Susan Sarandon, the CEO of British Petroleum, Leonardo DiCaprio's mom, and four-time Iditarod champion Lance Mackey, who carried one during the 2011 race. One young man who received a blue marble during a beach cleanup in Central California was inspired to turn his bar mitzvah into an "Ocean Mitzvah" at which he shared blue marbles with all the attendees.[3] A woman wrote that she had passed her marble on to her sister, who, rain or shine, had gone out every four weeks for twelve years to walk the beach at Pescadero, California, and record the health of that particular stretch of coastline. The marble helped the siblings have "a connecting moment": "We talk about everything but very seldom take a moment to really say 'Thanks for being a significant and positive influence on my little blue planet,'" she wrote.

One day in 2010, Andi Wong, technology instructor at the Rooftop School in San Francisco, called me after seeing a presentation to propose something. At the time, the kids in her school were enthralled by an educational video game called Marble Blast. Could we introduce the Blue Marbles Project to the school as a real-life version of the game? The school then created a curriculum that combined art, geography, biology, and ecology. Each child received a blue marble and was asked to pass it along to someone special. The class did research, and the kids picked people all over the world to whom they wanted to send marbles. They made boxes out of recycled maps, enclosed the marble and a note explaining the project, and sent the packages off.

It became an amazing project, as recipients got in touch with the class, sent photos of themselves with their blue marbles, and talked about who they planned to pass their marbles on to next. Marbles went to Paris, Berlin, Rome, the Sahara, New Zealand, Bolivia. A choral teacher took one to Cuba; a jazz musician took one to Tokyo. People photographed themselves and their marbles at the Monterey Bay Aquarium, on top of the Empire State Building, at the World Series. Marbles were given to kindergartners in China, to Hugh Jackman in San Francisco, to will.i.am in New York. It reminded the children how people and places are connected around the world, and how gratitude can make the world feel closer.

People ask me, "What do I have to do? What are the rules of the Blue Marble game?" The first rule is that the marble is blue. It doesn't matter where it comes from: some have been scavenged from childhood games, antiques shops, toy and craft stores. (You can also get marbles from our website www.bluemarbles.org.) The second rule is that you pass along the marble and thank the person you give it to, for what they do. The third is just a suggestion: that you share the Blue Marble story you created when you shared your marble, in any way that feels right—social media, poetry, song, a photograph, or best of all, spoken word. That's it. It's a fun, simple, and surprisingly moving experience based upon face-to-face gratitude. Whereas fines, policies, laws, monitoring, and research are part of the hard toolbox, blue marbles provide a complementary soft power to the movement to restore and protect our oceans and waterways.

When you get one, you'll understand. Then you have to give it away.

I have given away hundreds, thousands even (I always have two or three in my pockets, a tangible reminder with every step I take of water and gratitude), and it's always an interesting moment when you hand a marble to someone. For those receiving a marble, it can be unexpected, even uncomfortable. I've seen some people just look at it and say, "What the heck is *this?*" It can be a few moments before recipients really understand what the marble represents. But when they do, you can see that the gift stirs up something inside. And you can see them start to think of who they might like to give the marble to next.

I once presented blue marbles to people gathered for the International Association of Business Communicators, a group populated by vice presidents of strategic communication for a who's who list of Fortune 500 companies. If ever there was an audience of professionals with the potential for cynicism, I thought, this was it. At the end of my keynote address I asked attendees to hold up their blue marbles, and I explained why I had shared them. I suggested that they hold their marbles to their foreheads as they imagined the people they wanted to share them with (quite a sight from the podium, I assure you).

Then I asked them to hold their marbles to their hearts as they saw themselves passing along their marbles, imagining how this would make the recipient feel at that moment. There was a palpable but silent buzz in the room, and then something unexpected happened: a good number of those business-suited men and women began to tear up. We need to remember that we all deeply love our homes and each other—and when we take a moment to consider how very much we do, it can stir us at our core.

More recently I gave a marble to Dave Gomez, who served seventeen years of a life sentence for car theft until he was freed by a challenge to California's "three strikes" law. He looked at the marble in a way I'd never seen anyone do. "Free—that's what it feels to look at this marble," he told me.

What is it about a blue marble that can have such an effect? We know that, according to neuroscience, people like them because they're blue, and they're three-dimensional: they have weight (but not too much), temperature, texture (but not rough), mass, and we like pleasing, tangible objects. But a blue marble is really just a small piece of recycled glass, made of sand and a little cobalt to turn it blue. It costs a few pennies to manufacture. It's common, humble, plain, insignificant. And yet, when you look someone in the eyes and give them a blue marble with gratitude and intention, it can be the most important item in the world. Gratitude is a very powerful emotion: it opens people up, gets oxytocin flowing, and connects people to each other.

Each blue marble reminds us that we are connected to each other, emotionally and biologically. It symbolizes the deep connection between people and this planet that astronaut Eugene Cernan described as "the most beautiful star in the heavens—the most beautiful because it's the one we understand and we know, it's home, it's people, family, love, life—and besides that it is beautiful."[4] Looking at that small blue marble at the end of an outstretched arm invites a shift in perspective: we sometimes forget that from a million miles away we look so small and insignificant. Carl Sagan wrote of the Earth he described as a "pale blue dot" seen from a million miles away: "That's home. That's us. On it everyone you love, everyone you know, everyone you ever heard of, every human being who ever was, lived out their lives.... There is perhaps no better demonstration of the folly of human conceits than this distant image of our tiny world. To me, it underscores our responsibility to deal more kindly with one another and to preserve and cherish the pale blue dot, the only home we've ever known."[5]

Each blue marble tells us that everything we do on this planet matters. It represents clean, beautiful water that should be accessible to everyone. And not just water for drinking or growing crops: water to play in, jump in, splash in, throw at our friends; water to recreate in, to sail or surf or fish or swim or boat in; water to watch and listen to; water to go under and explore; water to love, cherish, care for, and protect. Our planet's waters are worth fighting for, risking everything for, and standing up for. Because being fully immersed in water, moving across or near water, really is everything you know it to be. It truly can do all that you think it can do for your body, your mind, and your relationships. Maybe that's why this simple idea is so easily ignored or shrugged off.

In a perfect world, our waters run healthy, clean, and free. Their waves and their currents and their stillness welcome all of us to heal, play, create, and love abundantly. For many of us, until that moment of observance or submergence, we work hard and struggle to maintain our ancient, personal connection to water. There is an interdependency with the natural world that goes beyond ecosystems, biodiversity, or economic benefits; our neurons and water need each other to live.

The revelation of this immensity . . . was like falling in love.

— GIORDANO BRUNO, FREETHINKING SIXTEENTH-CENTURY FRIAR, ON THE UNIVERSE[6]

All I really want to say is this:

Get in the water.

Walk along the water. Move across its surface. Get under it. Sit in it. Leap into it. Listen to it. Touch the water. Close your eyes and drink a big glass.

Fall more deeply in love with water in all its shapes, colors, and forms. Let it heal you and make you a better, stronger version of yourself. You need water. And water needs you now.

I wish you water.

Acknowledgments

Ultimately, the best way to forge a lasting relationship is to create something together. Whether it is a meal, an art project, or a spontaneous dance party, when you create something with others, you build a connection that lasts a lifetime.

— *The Social Synapse*, by Nora Epinephrine and Sarah Tonin

[This quote appears on the screen prior to shows by Blue Man Group at the Astor Place Theatre in New York City, where they have performed their explosive mix of music, color, and outrageous fun since 1991.]

I began thinking about Blue Mind as a nineteen-year-old sophomore at DePauw University in Greencastle, Indiana, when I met Barbara Daugherty. For eight months I spent Wednesday afternoons with Barbara at the nursing home she lived in. Sharing music with her opened my mind to the mysteries of the human brain. Thank you, Barbara.

I also owe a lot to my mentors, professors, and teachers at Barrington High School, DePauw University, Duke University, Northeastern University, and the University of Arizona who generously fed my curiosity about the natural world.

Thank you to my big, wild family for teaching me how to fall in love with water and piece myself together.

Thank you to the scientists, practitioners, artists, and

professionals who shared their research, stories, and insights through these pages. I so deeply admire your bright Blue Minds. All errors in translation are mine.

Thank you to the entire team at Little, Brown and Company. I've never worked with such an impeccably talented group. I am learning so much about books and writing from all of you. I hope that this is the book you'd imagined it would be.

Writing a book is like carving a block of wood. This one was like sculpting water. Thank you to my gifted editor, Geoff Shandler; literary agent, Wendy Keller; Blue Mind muse Vicki St. George; also Allie Sommer, Tim Whiting, Peggy Freudenthal, Chris Jerome, Elizabeth Garriga, Lisa Erikson, Michael Pietsch, Amanda Brown, Zoe Hood, Reagan Arthur, Andy Hine, Nathalie Morse, Heather Fain, Andy LeCount, Jin Yu, and Allison Warner, for providing the skilled hands needed for such an uncertain and unwieldy task. Let's all have a long swim together.

A special thank-you to Blue Mind interns Jasmin Aldridge, Emma Hazel, and Ruby Hoy.

In addition to those highlighted in the text, stimulating conversations about and steady support for Blue Mind while in, on, or near water with the following were invaluable in developing this work: Craig Adams, Patrick Adams, Chris Agnos, Abigail Alling, Barbara Andrews, Chris Andrews, Mark Beyeler, Jon Bowermaster, Fernando Bretos, Roberta Brett, Barbara Burgess, Halsey Burgund, Paulo Canas, Stuart Candy, Jenny Cavelle, Blayney Colmore, Fabien Cousteau, Don Croll, Larry Crowder, Serge Dedina, Carlos Delgado, Lance Descourouez, Rio Dluzak, Jake Dunagan, Tim Dykman, Elena Finkbeiner, Bryan Flores, Captain Joel Fogel, Ben Freiman, Rod Fujita,

Jennifer Galvin, Sharon Guynup, Nancy Higgs, Joshua Hoy, Jesús Ilundain, Pam Isom, Kiki Jenkins, Tony and Linda Kinninger, Sarah Kornfeld, Lorne Lanning, Harold Linde, Jeff Leifer, Jon Lowry, Kent Lowry, Jafari Mahtab, Amanda Martinez, Rod Mast, Matt McFadyen, David McGuire, Peter Mel, Cristina Mittermeier, Felix Moncada, Andy Myers, Peter Neill, Heather Newbold, Jack O'Neill, Tim O'Shea, Ran Ortner, Jennifer Palmer, Lindsay Peavey, Margot Pellegrino, Paul Piff, Nick Pilcher, Chris Pincetich, Aaron Pope, Jumal Qazi, Chris Reeves, Debra Reynolds, Michael Roberts, Grayson Roquemour, Anne Rowley, Jo Royle, Carl Safina, Rafe Sagarin, Shari Sant-Plummer, Roz Savage, Nik Sawe, Michael Scanlon, Darren Schreiber, Barton Seaver, Jesse Senko, Mark Shelley, Brian Skerry, Erik Soderholm, Mark Spalding, Julie Starke, Todd Steiner, Michael Stocker, Shane Toohey, Jim Toomey, Julia Townsend, Mark Van Thillo, Timothy Vogel, Colette Wabnitz, Scott Warren, Cameron Webb, Ben Wheeler, Mathew White, September Williams, and Steve Winter.

The four Blue Mind Summits are quite the characters in themselves and produced many new questions and discussions. Thank you to everyone who attended, sponsored, and helped organize these gatherings, including, but certainly not limited to, Petrina Bradbrook, Cory Crocker, Julia Davis, Max Davis, Kama Dean, Rio Dluzak, Justin DuClos, Sarah Kornfeld, Manuel Maqueda, Louise Nelson, Adam Norko, Ashleigh Papp, Amy Posey, Primavera Salva, Soni Singh, Kristen Weiss, and especially Julie Starke, who steadily shares her many talents and love of the ocean with the world.

Much gratitude to all of the blue angels who live like they love the ocean every day and provide support for our work.

Thank you to Alexi Murdoch for his song "Blue Mind":

Remember when you were only a child
And start to see with your blue mind
Start to see with your blue mind
Don't be afraid of what you find

Finally, thank you to the late Peter Matthiessen, who inspired me to write. "Nonfiction at its best is like fashioning a cabinet. It can never be sculpture. It can be elegant and very beautiful, but it can never be sculpture. Captive to facts—or predetermined forms—it cannot fly," he said.

Naturally, I think Blue Mind can swim.

Notes

Chapter 1: Why Do We Love Water So Much?

1. Robert Krulwich, "Born Wet, Human Babies Are 75 Percent Water. Then Comes the Drying," 26 November 2013, http://www.npr.org/blogs/krulwich/2013/11/25/247212488/born-wet-human-babies-are-75-percent-water-then-comes-drying.

2. On the face of it this is a simple poetic statement, but in fact it is quite a profound and transformative claim. Kurt Vonnegut, *Welcome to the Monkey House* (New York: Delacorte Press, 1968), xiii.

3. J. F. Helliwell, R. Layard, and J. Sachs, eds., *World Happiness Report 2013* (New York: UN Sustainable Development Solutions Network, 2013), 3.

4. Ibid., 4.

5. S. Lyubomirsky, K. M. Sheldon, and D. Schkade, "Pursuing happiness: the architecture of sustainable change," *Review of General Psychology* 9, no. 2 (2005), 111–31.

6. *World Happiness Report 2013,* 58.

7. J. H. Fowler and N. A. Christakis, "Dynamic spread of happiness in a large social network: longitudinal analysis over 20 years in the Framingham Heart Study," *British Medical Journal* 337, no. 2338 (5 December 2008), http://www.bmj.com/content/337/bmj.a2338.

8. *World Happiness Report 2013,* 69.

9. Madhu Kalia, "Assessing the Economic Impact of Stress—The Modern Day Hidden Epidemic," *Metabolism* 51, no. 6, Suppl. 1 (2002), 49–53.

Chapter 2: Water and the Brain: Neuroscience and Blue Mind

1. K. Smith, "Brain imaging: fMRI 2.0: Functional magnetic resonance imaging is growing from showy adolescence into a workhorse of brain imaging," *Nature* 484 (4 April 2012): 24–26, http://www.nature.com/news/brain-imaging-fmri-2-0-1.10365.

2. D. P. McCabe and A. D. Castel, "Seeing is believing: The effect of brain images on judgments of scientific reasoning," *Cognition* 107 (2008), 343–52, http://www.ncbi.nlm.nih.gov/pubmed/17803985.

3. Michael Posner, "Researchers develop 'camera' that will show your mind," *The Globe and Mail,* 4 February 2011, http://www.theglobeandmail.com/technology/science/researchers-develop-camera-that-will-show-your-mind/article565631.

4. A further development of EEG technology, magnetoencephalography (MEG), measures magnetic fields created by electrical activity in the brain to locate neuronal activity with greater precision. While the MEG machines are more like MRIs in that they are large, use powerful magnets, and must be shielded to avoid environmental interference, they are like EEGs in that the data they receive are not corrupted if the subject moves (unlike MRIs or fMRIs).

5. S. Satel and S. O. Lilienfeld, *Brainwashed: The Seductive Appeal of Mindless Neuroscience* (New York: Basic Books, 2013), 38.

6. A. Wright, "Brain Scanning Techniques (CT, MRI, fMRI, PET, SPECT, DTI, DOT)," white paper (Carmarthen, U.K.: Cerebra, 2010), http://www.cerebra.org.uk/SiteCollectionDocuments/Research%20PDF%27s/Brain%20scanning%20techniques.pdf.

7. In 2010, neuroscientists used fMRI in more than 1,500 published articles, while the number of papers using PET or SPECT scans leveled off between 2000 and 2010. Papers citing data gathered by EEG or MEG also increased, but the total was more than a third below that of fMRI. See "Brain imaging: fMRI 2.0," by Kerri Smith, *Nature,* 484, no. 7392 (4 April 2012), http:// www.nature.com/news/brain-imaging-fmri-2-0-1.10365.

8. This description is based on the article "MRI brain scan: what it feels like," by Anni Sofferet, http://voices.yahoo.com/mri-brain-scan-feels-like-7644620.html, and "What does having an fMRI scan involve?" from the University of Manchester website, http://www.bbmh.manchester.ac.uk/resilience/fMRI/fMRIrisks.

9. Wright, "Brain Scanning Techniques." 8–9.

10. Lizzie Buchen, "Neuroscience Illuminating the Brain," http://www.nature.com/news/2010/100505/full/465026a.html. Today, some of the most exciting neuroscience research is being done by using optogenics to study neural circuitry in the brains of animals.

11. Cambridge Research Systems has developed two devices based on fNIRS (functional near-infrared spectroscopy), which measures hemodynamic changes associated with neuron behavior. See http://www.crsltd.com/tools-for-functional-imaging/tomographic-fnirs-imaging/nirscout.

12. Kenneth Saladin, *Anatomy and Physiology: The Unity of Form and Function* (New York: McGraw-Hill, 2007), 520. For an excellent essay on the distraction of discursive thought, see "We are lost in thought" by neuroscientist Sam Harris, in *This Will Make You Smarter: New Scientific Concepts to Improve Your Thinking*, John Brockman, ed. (New York: Harper Perennial, 2012), 211–13.

13. Nikhil Swaminathan, "Why does the brain need so much power?" *Scientific American,* 29 April 2008, http://www.scientificamerican.com/article.cfm?id=why-does-the-brain-need /s. Based on the study "Tightly coupled brain activity and cerebral ATP metabolic rate," by Fei Du, Xiao-Hong Zhu, Yi Zhang, Michael Friedman, Nanyin Zhang, Kamil Ugurbil, and Wei Chen. *Proceedings of the National Academy of Sciences* 105, no. 17 (2008), 6409–14; published ahead of print 28 April 2008, doi:10.1073/pnas.0710766105.

14. The figure of 100 billion was challenged in 2009 by a researcher in Brazil. Dr. Suzana Herculano-Houzel and her team took the brains of four adult men who had died of nonneurological diseases, combined them, and dissolved the cell membranes within the brains. She took a sample of this "brain soup," counted the cell nuclei of the neurons present, and then extrapolated the total from there. On average, she found, the human brain contains 86 billion neurons, 14 billion fewer than the previous estimate. She also discovered that the proportion of nonneuronal cells to neurons (estimated to be as much as 10 to 1), was actually close to equal: 86.1 neurons to 84.6 nonneuronal cells. See "Equal numbers of neuronal and nonneuronal cells make the human brain an isometrically scaled-up primate brain," by Frederico A. C. Azevedo, Ludmila R. B. Carvalho, Lea T. Grinberg, José Marcelo Farfel, Renata E. L. Ferretti, Renata E. P. Leite, Wilson Jacob Filho, Roberto Lent, Suzana Herculano-Houzel. *Journal of Comparative Neurology* 513, no. 5 (2009), 532–41.

15. Mahzarin Banaji, "A solution for collapsed thinking: signal detection theory," in *This Will Make You Smarter,* 389–93.

16. See the website for the Human Connectome Project, http://www.humanconnectomeproject.org. See also Marco Iacoboni, "Like attracts like," in *This Will Make You Smarter,* 330–32.

17. See S. Herculano-Houzel, "The human brain in numbers: a linearly scaled-up primate brain," *Frontiers of Human Neuroscience* 3 (9 November 2009), 31, http://www.ncbi.nlm.nih.gov/pubmed/19915731.

18. John Medina, *Brain Rules: 12 Principles for Surviving and Thriving at Work, Home, and School* (Seattle: Pear Press, 2008), 31–32.

19. Studies show that mammalian brains normally lose about half of their synapses from childhood to puberty. See G. Chechik, I. Mellijson, and E. Ruppin, "Neuronal regulation: a mechanism for synaptic pruning during

brain maturation," *Neural Computation* 11, no. 8 (15 November 1999), 2061–80, http://www.ncbi.nlm.nih.gov/pubmed/10578044.

20. Mark Changzi, *Harnessed: How Language and Music Mimicked Nature and Transformed Ape to Man* (Dallas: BenBella Books, 2011), see page 27 for a diagram.

21. David Pizarro, "Everyday Apophenia," in *This Will Make You Smarter,* 394.

22. For an interesting discussion of perception and whether the brain gathers data from the senses and then uses them to make a prediction ("bottom-up" processing) or compares data against predictive models already stored in the brain ("top-down" processing), see Andy Clark, "Predictive Coding," in *This Will Make You Smarter,* 132–34.

23. Gerald Smallberg, neurologist, "Bias Is the Nose for the Story," in *This Will Make You Smarter,* 43–45. There are also fascinating studies of how the brain "sees" a baseball before it can truly perceive it. See Y. Anwar, "Hit a 95 mph baseball? Scientists pintpoint how we see it coming," UC Berkeley Newscenter, 8 May 2013, http://newscenter.berkeley.edu/2013/05/08/motion-vision. The study referenced is G. W. Maus, J. Fischer, and D. Whitney, "Motion-dependent representation of space in area MT+," *Neuron* 78, no. 3 (8 May 2013), 554–62, http://www.cell.com/neuron/abstract/S0896-6273%2813%2900257-2.

24. David Eagleman, *Incognito: The Secret Lives of the Brain* (New York: Pantheon, 2011), 22.

25. Norman Doidge writes in *The Brain That Changes Itself: Stories of Personal Triumph from the Frontiers of Brain Science* (New York: Penguin, 2007), "Merzenich.... found that as neurons are trained and become more efficient, they can process faster. This means that the speed at which we think is itself plastic. Speed of thought is essential to our survival" (67). Also see M. Merzenich, *Soft Wired* (San Francisco: Parnassus, 2013).

26. M. Quallo, C. Price, K. Ueno, T. Asamizuya, K. Cheng, R. Lemon, and A. Iriki, "Gray and White Matter Changes Associated with Tool-Use Learning in Macaque Monkeys," *Proceedings of the National Academy of Sciences* 106, no. 43 (27 October 2009), 18379–84, http:// www.pnas.org/ content/106/43/18379. A well-known study of London taxi drivers showed that their hippocampi (the region in the brain involved in memory formation) were physically larger than in non-taxi-driving individuals of a similar age. E. A. Maguire, D. G. Gadian, I. S. Johnsrude, C. D. Good, J. Ashburner, R. S. Frackowiak, and C. D. Frith, "Navigation-related structural change in the hippocampi of taxi drivers," *Proceedings of the National Academy of Sciences* 97, no. 8 (27 April 2000), 4398–4403, http://www.pnas.org/content/97/8/4398.full.

27. L. Jiang, H. Xu, and C. Yu, "Brain connectivity plasticity in the motor network after ischemic stroke," *Neural Plasticity,* 2013, article ID 924192, http://www.hindawi.com/journals/np/2013/924192.

28. In *Brain Rules,* John Medina cites the example of a neurosurgeon who combined the brain maps of 117 of his patients, to discover that there was only one region of the brain where even 79 percent had a critical language area. Medina writes that before operating, "[The surgeon] has to map each individual's critical function areas because *he doesn't know where they are*" (65–66).

Chapter 3: The Water Premium

1. Details of the 2003 Coastal Trail Expedition can be found at http://www.californiacoastaltrail.info.

2. "Neighbors dig deep in bidding war over narrow strip of real estate," *AOL Real Estate,* 5 September 2013, http://realestate.aol.com/blog/writers/aol-real-estate-editors/rss.xml.

3. G. Bachelard, *Water and Dreams: An Essay on the Imagination of Matter,* trans. E. R. Farrell (Dallas: The Dallas Institute of Humanities and Culture, 1983), 32–33.

4. D. G. Myers and E. Diener, "Who is happy?," *Psychological Science* 6 (1995), 10–19.

5. R. T. Howell, "How many happy balls are in your beaker?," *Psychology Today,* 31 July 2012, http://www.psychologytoday.com/blog/cant-buy-happiness/201207/how-many-happy-balls-are-in-your-beaker.

6. *World Happiness Report 2013,* 3 (see chap. 1, n. 2).

7. J. P. Forgas, "When sad is better than happy: Negative affect can improve the quality and effectiveness of persuasive messages and social influence strategies," *Journal of Experimental Social Psychology* 43 (2007), 513–28, www.sciencedirect.com/science/article/pii/S0022103106000850.

8. L. Kováč, "The biology of happiness: chasing pleasure and human destiny," *European Molecular Biology Organization (EMBO) Reports* 13, no. 4 (April 2012), 297–302, http://www.ncbi.nlm.nih.gov/pmc/articles/PMC3321158/#!po=5.55556.

9. B. L. Fredrickson, K. M. Grewen, K. A. Coffey, S. B. Algoe, A. M. Firestine, J. M. G. Arevalo, J. Ma, and S. W. Cole, "A functional genomic perspective on human well-being," *Proceedings of the National Academy of Sciences,* published online before print 29 July 2013, www.pnas.org/cgi/doi/10.1073/pnas.1305419110.

10. S. Lyubomirsky, K. M. Sheldon, and D. Schkade, "Pursuing happiness: the architecture of sustainable change," *Review of General Psychology* 9, no. 2 (2005), 111–31.

11. Kováč, "The biology of happiness," 5.

12. Barbara L. Fredrickson, Karen M. Grewen, Kimberly A. Coffey, Sara B. Algoe, Ann M. Firestine, Jesusa M. G. Arevalo, Jeffrey Ma, and Steven W.

Cole, "A functional genomic perspective on human well-being," *Proceedings of the National Academy of Sciences*, published online before print 29 July 2013, doi:10.1073/pnas.1305419110.

13. M. A. Max-Neef, *Human Scale Development: Conception, Application and Further Reflections* (New York and London: Apex Press, 1991), 32–33.

14. Winifred Gallagher, *The Power of Place: How Our Surroundings Shape Our Thoughts, Emotions, and Actions* (New York: Poseidon, 1993), 125.

15. Daniel J. Siegel, *Mindsight: The New Science of Personal Transformation* (New York: Bantam Books, 2011), 111.

16. Sally Satel and Scott O. Lilienfeld write, "Fibrous pathways run from the frontal lobes, which are associated with impulse control and risk assessment, to the amygdala, which is linked to the primitive impulses of aggression, anger, and fear, among other emotions. Optimally, the frontal lobes modulate the amygdala, a working relationship that depends on a well-functioning connection between the two" (*Brainwashed*, 99; see chap. 2, n. 5).

17. For a discussion of the function of dopamine in addiction, see chapter 6.

18. M. E. Raichle, A. M. MacLeod, A. Z. Snyder, W. J. Powers, A. D. Gusnard, and G. L. Shumlan, "A default mode of brain function," *Proceedings of the National Academy of Sciences* 98, no. 2 (January 2001), 676–82.

19. These "flashbulb" memories are often a feature of PTSD and can be difficult to erase. See chapter 5; also B. M. Law, "Seared in our memories," *Monitor on Psychology* 42, no. 8 (September 2011), 60, http://www.apa.org/monitor/2011/09/memories.aspx.

20. For an excellent discussion of how emotions are involved in almost every decision we make, see Terrence J. Sejnowski, "Nature Is More Clever Than We Are," in *This Explains Everything*, John Brockman, ed. (New York: Harper Perennial, 2013), 328–31.

21. Zajonc, R. B. "Feeling and Thinking: Preferences need no inferences." *American Psychologist* 35, no. 2 (February 1980), 151–75.

22. G. W. Kim, G. W. Jeong, T. H. Kim, H. S. Baek, S. Oh, H. Kang, S. W. Lee, Y. S. Kim, and J. K. Song, "Functional neuroanatomy associated with natural and urban scenic views in the human brain: 3.0T functional MR imaging." *Korean Journal of Radiology* 11, no. 5 (September–October 2010), 507–13, http://www.ncbi.nlm.nih.gov/pmc/articles/PMC2930158.

23. T. H. Kim, G. W. Jeong, H. S. Baek, G. W. Kim, T. Sundaram, H. K. Kang, S. W. Lee, H. J. Kim, and J. K. Song, "Human brain activation in response to visual stimulation with rural and urban scenery pictures: a functional magnetic resonance imaging study." *The Science of the Total Environment* 408, no. 12 (15 May 2010), 2600-2607, http://www.ncbi.nlm.nih.gov/pubmed/20299076.

24. X. Yue, E. A. Vessel, and I. Biederman, "The neural basis of scene preferences," *Neuroreport* 18, no. 6 (16 April 2007), 525-29, http://www.ncbi.nlm.nih.gov/pubmed/17413651.

25. M. White, A. Smith, K. Humphreys, S. Pahl, D. Snelling, and M. Depledge, "Blue space: the importance of water for preference, affect, and restorativeness ratings of natural and built scenes," *Journal of Environmental Psychology* 30, no. 4 (2010), 482–93.

26. M. Gross, "Can science relate to our emotions?" *Current Biology* 23, no. 12 (17 June 2013), R501–4, http://www.cell.com/current-biology/fulltext/S0960-9822%2813%2900681-7.

27. K. S. Kassam, A. R. Markey, V. L. Cherkassky, G. Lowenstein, and M. A. Just, "Identifying emotions on the basis of neural activation," *PLoS ONE* 8, no. 6 (2013), e66032, doi:10.1371/journal.pone.0066032, http://www.plosone.org/article/info%3Adoi%2F10.1371%2Fjournal.pone.0066032.

28. R. Hanson, *Hardwiring Happiness: The New Brain Science of Contentment, Calm, and Confidence* (New York: Crown, 2013), 10.

29. W. A. Cunningham and T. Kirkland, "The joyful, yet balanced, amygdala: moderated responses to positive but not negative stimuli in trait happiness," *Social Cognitive and Affective Neuroscience* (18 May 2013), epub ahead of print, doi: 10.1093/scan/nst045, http://scan.oxfordjournals.org/content/early/2013/05/17/scan.nst045.abstract.

30. Loretta Graziano Breuning, "Are You Addicted to Empathy?," *Psychology Today*, 17 October 2013, http://www.psychologytoday.com/blog/your-neurochemical-self/201310/are-you-addicted-empathy.

31. Kim et al., "Functional neuroanatomy," 507–13, http://www.ncbi.nlm.nih.gov/pmc/articles/PMC2930158/#__ffn_sectitle.

32. G. MacKerron and S. Mourato, "Happiness is greater in natural environments," *Global Environmental Change* 23, no. 5 (October 2013), 992–1000, http://www.sciencedirect.com/science/article/pii/S0959378013000575.

33. A. W. Vemuri and R. Costanza, "The role of human, social, built, and natural capital in explaining life satisfaction at the country level: toward a National Well-Being Index (NWI)," *Ecological Economics* 58 (2006), 119–33, http://www.sciencedirect.com/science/article/pii/S092180090500279X.

34. J. M. Zelenski and E. K. Nisbet, "Happiness and feeling connected: the distinct role of nature relatedness," *Environment and Behavior* 46, no. 1 (January 2014), 3–23, http://eab.sagepub.com/content/46/1/3.abstract. Published online before print, doi: 10.1177/0013916512451901.

35. Elise Proulx, "3 insights from the frontiers of positive psychology," *Huffington Post*, 13 August 2013, http://www.huffingtonpost.com/2013/08/13/positive-psychology-insights_n_3745523.html?view=print&comm_ref=false.

36. C. O'Brien, "A footprint of delight: exploring sustainable happiness," *NCBW Forum* (1 October 2006), 1–12, www.bikewalk.org/pdfs/forumarch1006 footprint.pdf.

37. M. Gross, "Can science relate to our emotions?"

38. MacKerron and Mourato, "Happiness is greater," 6.

39. J. Pretty, J. Peacock, R. Hine, M. Sellens, N. South, and M. Griffin, "Green exercise in the UK countryside: effects on health and psychological well-being, and implications for policy and planning," *Journal of Environmental Planning and Management* 50, no. 2 (March 2007), 211–31, www.greenexercise .org/pdf/JEPM%20-%20CRN%20Study.pdf.

40. MacKerron and Mourato, "Happiness is greater," 7.

41. Ibid., 3.

42. F. Brereton, J. P. Clinch, and S. Ferreira, "Happiness, geography, and the environment," *Ecological Economics* 65, no. 2 (1 April 2008), 386–96, http://www.sciencedirect.com/science/article/pii/S0921800907003977.

43. J. Barton and J. Pretty, "What is the best dose of nature and green exercise for improving mental health? A multi-study analysis," *Environmental Science & Technology* 44 (May 2010), 3947–55, http://www.ncbi.nlm.nih.gov/ pubmed/20337470.

44. S. Volker and T. Kistemann, "'I'm always entirely happy when I'm here!' Urban blue enhancing human health and well being in Cologne and Düsseldorf, Germany," *Social Science & Medicine* 91 (August 2013), 141–52, http://www.ncbi.nlm.nih.gov/pubmed/23273410.

45. According to the official website, San Antonio's River Walk, or Paseo del Rio, is the top tourist destination in the state of Texas, and the largest urban ecosystem in the nation. Steps away from the Alamo, it provides "a serene and pleasant way to navigate the city," http://www.thesanantonioriv erwalk.com.

46. U.S. Census Bureau, 2011. Census 2010, http://factfinder2.census .gov/faces/nav/jsf/pages/index.xhtml.

47. "Neuroeconomics: The Cost of Love," *Duke Magazine*, 25 July 2013, http://dukemagazine.duke.edu/article/neuroeconomics-the-cost-of-love.

48. Leaf Van Boven and Thomas Gilovich, "To Do or to Have? That is the Question," *Journal of Personality and Social Psychology* 85, no. 6 (2003), 1193–1202.

49. NOAA's State of the Coast website, U.S. Population Living in Coastal Watershed Counties, http://stateofthecoast.noaa.gov/population/welcome .html.

50. Elise Proulx, "3 Insights."

51. Yi-Fu Tuan, *Space and Place: The Perspective of Experience* (Minneapolis: University of Minnesota Press, 1976; reprint 2001). Notes from

Amazon.com book description, http://www.amazon.com/Space-Place-The-Perspective-Experience/dp/0816638772/ref=sr_1_1?ie=UTF8&qid=1387149668&sr=8-1&keywords=space+and+place+the+perspective+of+experience.

52. Other factors include value of nearby properties and the size and shape of the lot. "ECU Prof Studies Value of Beach Property," East Carolina University press release, 20 September 1993, http://www.ecu.edu/cs-admin/news/newsstory.cfm?ID=31.

53. Les Christie, "Floating Homes: What It Costs to Live on the Water," *CNN Money*, 15 June 2012, http://realestate.aol.com/blog/2012/06/15/cnn-floating-home.

54. M. J. Seiler, M. T. Bond, and V. L. Seiler, "The Impact of World Class Great Lakes Water Views on Residential Property Values," *Appraisal Journal* 69, no. 3 (2001), 287–95.

55. Avalon is a sad example of how expensive oceanfront properties can *end up*: the town's beaches, and many of its homes, were destroyed by Hurricane Sandy in 2012.

56. Christopher Major and Kenneth Lusht, "The Beach Study: An Empirical Analysis of the Distribution of Coastal Property Values," Department of Insurance and Real Estate, Smeal College of Business, Pennsylvania State University, http://www.gradschool.psu.edu/index.cfm/diversity/mcnair/papers2003/majorpdf.

57. S. Yu, S. Han, and C. Chai, "Modeling the Value of View in Real Estate Valuation: A 3-D GIS Approach," working paper 2004, http://prres.net/Papers/Yu_Modelling_The_Value_Of_View.Pdf.

58. Courtney Trenwith, "Palm Jumeirah Property Prices Soar in Q2," *Arabian Business*, 4 July 2013, http://www.arabianbusiness.com/palm-jumeirah-property-prices-soar-in-q2-507755.html.

59. Brian Milligan, "How Much Should You Pay for a Sea View?," *BBC News*, 11 July 2013, http://www.bbc.co.uk/news/business-23255452.

60. J. Luttik, "The Value of Trees, Water and Open Space as Reflected by House Prices in the Netherlands," *Landscape and Urban Planning* 48, no. 3–4 (May 2000), 161–67.

61. V. Grigoriadis, "Bohemian Cove," *Vanity Fair*, March 2011, http://www.vanityfair.com/hollywood/features/2011/03/paradise-cove-201103.

62. *The Expedia 2013 Flip Flop Report* surveyed preferences among 8,606 beachgoers in North America, South America, Europe, Asia, and Australia, http://viewfinder.expedia.com/news/expedia-2013-flip-flop-report?brandcid=social.vf.Expedia%20Guest%20Author.Features.d60a33a7-50ee-63e1-bbde-ff000073f150.

63. The number of cruise ship passengers forecast for 2011 source: Cruise Industry Overview 2013: The State of the Cruise Industry, the Florida-Caribbean Cruise Association, http://www.f-cca.com/downloads/2013-cruise-industry-overview.pdf.

64. White et al., "Blue space."

65. Ekart Lange and Peter V. Schaeffer, "A Comment on the Market Value of a Room with a View," *Landscape and Urban Planning* 55, no. 2 (2001), 113–20.

66. M. White, "Health and Well Being from Coastal Environments," presented to Delivering Sustainable Coasts, SUSTAIN International Conference, Southport (UK), 18–19 September 2012, 2, http://www.sustain-eu.net/news/visits/sefton/presentations/white.pdf.

67. Hillary Huffer, "The Economic Value of Resilient Coastal Communities," NOAA draft report, 18 March 2013, 2, http://www.ppi.noaa.gov/wp-content/uploads/EconomicValueofResilientCoastalCommunities.pdf.

68. National Ocean Economics Program 2009, State of the U.S. Ocean and Coastal Economies, http://www.oceaneconomics.org/NationalReport.

69. Economists make a clear distinction between market (pharmaceuticals) and nonmarket (a view and its associated stress reduction) goods and services. While our current understanding using ecosystem service valuation methods is incomplete, nonmarket value must—and will—be included in policy- and decision-making processes.

70. "Europe Floods 2013's Costliest Natural Disaster," Associated Press, 9 July 2013, http://www.weather.com/news/europe-floods-2013s-costliest-natural-disaster-20130709.

71. According to the New England Center for Investigative Reporting, over the past four decades of destruction and repair the house at 48 Oceanside Drive, Scituate, Massachusetts (south of Boston)—with a great view of stunning beaches and dramatic cliffs—has been damaged at least nine times and the federal flood insurance program has spent upwards of $750,000 to rebuild it, http://necir.org.

72. M. White, S. Pahl, K. Ashbullby, S. Herbert, and M. Depledge, "Feelings of Restoration from Recent Nature Visits," *Journal of Environmental Psychology* 35 (September 2013), 40–51. Interestingly, in a 2011 study researchers found that restorativeness experienced from visiting coastal areas was heightened when (1) the ambient temperature was cooler than the monthly average for the area, and (2) the water and air quality were better than normal. See J. A. Hipp and O. A. Ogunseitan, "Effect of Environmental Conditions on Perceived Psychological Restorativeness of Coastal Parks," *Journal of Environmental Psychology* 31, no. 4 (2011), 421–29.

73. In 2008 a couple in New Jersey sued the state for building a protective sand dune between their house and the water that blocked their ocean view. The jury in the case awarded them $375,000 in compensation for the lost value of their property—even though the protective dune saved their $2 million home from being destroyed by Superstorm Sandy in October 2012. (In 2013 the verdict was overturned by the New Jersey Supreme Court.)

74. http://www.dosomething.org/tipsandtools/11-facts-about-bp-oil-spill.

75. Suzanne Goldenberg, "US Government Assessment of BP Oil Spill 'Will Not Account for Damage,'" *The Guardian*, 11 July 2013, http://www.theguardian.com/environment/2013/jul/11/us-assessment-bp-oil-spill-damage.

76. "BP Oil Spill Caused Significant Psychological Impact Even to Nearby Communities Not Directly Touched by Oil," University of Maryland School of Medicine press release, 17 February 2011, http://umm.edu/news-and-events/news-releases/2011/bp-oil-spill-caused-significant-psychological-impact-even-to-nearby-communities-not-directly-touched-by-oil#ixzz2fHgTriAd.

77. Samuel L. Clemens, *Life on the Mississippi* (Boston: H. O. Houghton, 1883).

Chapter 4: The Senses, the Body, and "Big Blue"

1. Scientists now state that we have anywhere from ten to thirty different senses, including temperature, the vestibular sense (allowing us to sense our spatial orientation, as well as its direction, and speed), a sense of time, a sense of intuition (some researchers believe this is actually the ability to sense danger in the environment), and, according to recent research, the ability to discern quantities visually—a "number sense" (B. M. Harvey, B. P. Klein, N. Petridou, and S. O. Dumoulin, "Topographic Representation of Numerosity in the Human Parietal Cortex," *Science* 341, no. 6150 (6 September 2013), 1123–26, http://www.sciencemag.org/content/341/6150/1123.abstract.). Many of those thirty senses, however, have to do with perceiving the internal world—blood sugar levels, respiratory rate, hunger, thirst, pain, the gag reflex, and even the sense of fullness in the bladder or colon.

2. Byte quantities differ depending on whether one is talking about disk storage or processor storage (the later including twenty-four more of the smaller component).

3. See Marcus E. Raichle,"The Brain's Dark Energy," *Scientific American*, March 2010, 44–49. Your brain can fill in missing information that your senses can't perceive; for example, you never see the "blind spots" in your field of vision in both your left and right eye because your brain automatically fills in the necessary visual data.

4. There's a specific location in the brain, the inferior parietal lobule (IPL), at the intersection of the occipital (vision), temporal (hearing), and parietal (touch) lobes, that's designed to receive, process, and integrate sensory data in such a way that the world makes sense.

5. Abram, *The Spell of the Sensuous: Perception and Language in a More-Than-Human World* (New York: Vintage Books, 1996), 60.

6. In "The Brain's Dark Energy," Dr. Raichle theorizes that, because so little visual information is used to formulate our conscious perceptions (100 bits per second), "the brain probably makes constant predictions about the outside environment in anticipation of paltry sensory inputs reaching it from the outside world" (47). There is also an excellent essay on how the brain creates perceptual maps from sensory data by physicist Frank Wilczek: "Hidden Layers," in Brockman, *This Will Make You Smarter*, 188–91.

7. Gerald Smallberg, "Bias Is the Nose for the Story," in Brockman, *This Will Make You Smarter*, 43.

8. While these individuals could see color, they had no idea what those colors were, and had to be taught. For an excellent description of the experience of gaining vision after years of blindness, see "Into the Light" by Robert Kurson, *Esquire*, June 2005, http://www.esquire.com/features/ESQ0605BLIND_114. See also "Vision Following Extended Congenital Blindness," by Yuri Ostrovsky, Aaron Andalman, and Pawan Sinha, *Psychological Science* 17, no. 12 (2006), 1009–14.

9. Eagleman, *Incognito*, p. 41 (see chap. 2, n. 25). See also "The blind climber who 'sees' with his tongue," by Buddy Levy, *Discover*, 23 June 2008, http://discovermagazine.com/2008/jul/23-the-blind-climber-who-sees-through-his-tongue.

10. Laura Sewall, "The skill of ecological perception," in *Ecopsychology: Restoring the Earth, Healing the Mind*, Theodore Roszak, Mary E. Gomes, and Allen D. Kanner, eds. (San Francisco: Sierra Club Books, 1995), 206.

11. "Brain Plasticity," http://merzenich.positscience.com/about-brain-plasticity.

12. Abram, *The Spell of the Sensuous*, ix.

13. While in the future we all may be able to immerse ourselves in a complete *Star Trek* "holodeck" kind of virtual experience, in today's virtual reality environments many of the senses we rely upon in the real world are not stimulated to the same degree. In a 2013 study, researchers in Los Angeles compared place cells (which indicate our position in space) active in rats running in an immersive virtual reality system versus the real world. Twice as many neurons in place cells were active in the real world as in virtual reality. The researchers theorized that "vestibular and other sensory cues present in

[the real world] are necessary to fully activate the place-cell population." P. Ravassard, A. Kees, B. Williers, D. Ho, D. Aharoni, J. Cushman, Z. M. Aghajan, and M. R. Mehta, "Multisensory control of hippocampal spatiotemporal selectivity," *Science* 340, no. 6138 (14 June 2013), 1342–46.

14. V. S. Ramachandran, *The Tell-Tale Brain*, Kindle locations 1245–59 and 1280–1303.

15. Hayley Dixon, "Blue Lagoon dyed black to deter swimmers," *The Telegraph*, 11 June 2013, http://www.telegraph.co.uk/earth/10112837/Blue-Lagoon-dyed-black-to-deter-swimmers.html.

16. Natalie Angier, "True Blue Stands Out in an Earthly Crowd," *New York Times*, 23 October 2012, http://www.nytimes.com/2012/10/23/science/with-new-findings-scientists-are-captivated-by-the-color-blue.html?pagewanted=all&_r=1&. Also "Are Some Things Universally Beautiful? Part 1 of the TED Radio Hour episode What Is Beauty," 19 April 2013, http://m.npr.org/story/174726813.

17. "True Colors—Breakdown of Color Preferences by Gender," http://blog.kissmetrics.com/gender-and-color.

18. Leo Widrich, "Why is Facebook blue? The science behind colors in marketing," *Fast Company*, 16 May 2013, http://www.fastcompany.com/3009317/why-is-facebook-blue-the-science-behind-colors-in-marketing.

19. Angier, "True Blue."

20. Susana Martinez-Conde, "The Color of Pain," *Scientific American*, 15 August 2013, http://blogs.scientificamerican.com/illusion-chasers/2013/08/15/the-color-of-pain/ - respond.

21. G. Vandewalle, S. Schwartz, D. Grandjean, C. Wuillaume, E. Balteau, C. Degueldre, M. Schabus, C. Phillips, A. Luxen, D. J. Dijk, and P. Maquet, "Spectral quality of light modulates emotional brain responses in humans," *Proceedings of the National Academy of Sciences*, 24 September 2010, http://www.pnas.org/content/early/2010/10/14/1010180107.abstract.

22. Angier, "True Blue."

23. "Different colors describe happiness and depression," MSNBC LiveScience, 8 February 2010, http://www.msnbc.msn.com/id/35304133/ns/technology _and_science-science.

24. Adam Alter, *Drunk Tank Pink, and Other Unexpected Forces That Shape How We Think, Feel, and Behave* (New York: Penguin, 2013), 157–58.

25. Vanderwalle et al., "Spectral quality," 24 September 2010, www.pnas.org/cgi/doi/10.1073/pnas.1010180107.

26. Christian Jarrett, "Colors affect mental performance, with blue boosting creativity," http://www.bps-research-digest.blogspot.co.uk/2009/02/colours-affect-mental-performance-with.html#.URE_ahI018I.twitter. Study

cited: Ravi Mehta, Rui (Juliet) Zhu, "Blue or Red? Exploring the Effect of Color on Cognitive Task Performances," *Science* 323, no. 5918 (27 February 2009), 1226–29.

27. In 1996 Eric Charlesworth reviewed *BLUE Magazine* in the October–November issue of *Inside Media*. Alas, the publication was slightly ahead of the wave and published its final issue in March 2000, http://www.bluemagazine.com.

28. R. G. Coss, S. Ruff, T. Simms, "All that glistens II: the effects of reflective surface finishes on the mouthing activity of infants and toddlers," *Ecological Psychology* 15, no. 3 (2003), 197–213.

29. Bachelard, *Water and Dreams*, 145 (see chap. 3, n. 3).

30. K. Abe, K. Ozawa, Y. Suzuki, and T. Sone, "The effects of visual information on the impression of environmental sounds," *Inter-Noise 99*, 1177–82. Cited in Josh H. McDermott, "Auditory Preferences and Aesthetics: Music, Voices, and Everyday Sounds," in *Neuroscience of Preference and Choice: Cognitive and Neural Mechanisms,* Raymond J. Dolan and Tali Sharot, eds. (London: Elsevier Press, 2012).

31. A SmartPlanet interview with David Z. Hambrick, associate professor of psychology at Michigan State University, reminded me that television shows, 3-D movies, and smartphone apps are unlikely to replace the cognitive and emotional values and services provided by being outside in nature. Christie Nicholson, "New evidence fails to replicate the very study upon which the brain-training game industry depends," 28 May 2012, http://www.smartplanet.com/blog/thinking-tech/qa-new-evidence-shows-brain-training-games-dont-work.

32. Charles Fishman, *The Big Thirst: The Secret Life and Turbulent Future of Water* (New York: Free Press, 2012), 311.

33. K. A. Rose, I. G. Morgan, J. Ip, et al., "Outdoor Activity Reduces the Prevalence of Myopia in Children," *Ophthalmology* 115, no. 8 (2008), 1279–85.

34. There is no such thing as water with no taste and smell, but most water providers are very aware that their customers expect their water to taste and smell "good." There's even a foundation (the Water Research Foundation, formerly AwwaRF) "dedicated to advancing the science of water by sponsoring cutting-edge research and promoting collaboration." See "Advancing the Science of Water: AwwaRF and Research on Taste and Odor in Drinking Water," 2007, http://www.waterrf.org/resources/StateOfTheScienceReports/TasteandOdorResearch.pdf.

35. See Esther Inglis-Arkell, "What really causes that amazing after the rain smell?," 20 August 2013, http://io9.com/what-really-causes-that-amazing-after-the-rain-smell-1167869568; Andrea Thompson, "Key found to the smell of the sea," 1 February 2007, http://www.livescience.com/4313-key

-smell-sea.html, and Katharina D. Six, Silvia Kloster, Tatiana Ilyina, Stephen D. Archer, Kai Zhang, and Ernst Maier-Reimer, "Global warming amplified by reduced sulfur fluxes as a result of ocean acidification," *Nature Climate Change* 2013, published 25 August 2013, doi:10.1038/nclimate1981, http://www.nature.com/nclimate/journal/vaop/ncurrent/full/nclimate1981.html.

36. "Cognitive Facilitation Following Intentional Odor Exposure," *Sensors* 11 (2011), 5469–88, doi:10.3390/s110505469.

37. J. Lehrner, G. Marwinski, P Johren, and L. Deecke, "Ambient odors of orange and lavender reduce anxiety and improve mood in a dental office," *Physiology & Behavior* 86, nos. 1–2 (15 September 2005), 92–95.

38. In "How Taste Works," Sarah Dowdey writes, "Chemical stimuli activate the chemoreceptors responsible for gustatory and olfactory perceptions. Because taste and smell are both reactions to the chemical makeup of solutions, the two senses are closely related. If you've ever had a cold during Thanksgiving dinner, you know that all of the subtlety of taste is lost without smell." http://science.howstuffworks.com/life/human-biology/taste.htm.

39. Charles Spence, "Auditory contributions to flavour perception and feeding behaviour," *Physiology & Behavior* 107, no. 4 (5 November 2012), 505–15, doi: 10.1016/j.physbeh.2012.04.022. Epub 2012 May 2.

40. Francesca Bacci and David Melcher, eds., "Sound bites: how sound can affect taste," excerpted from *Art and the Senses* (Oxford: Oxford University Press, 2011), http://blog.oup.com/2011/08/sound-bites.

41. A hint to Japanese philosophy where the five elements are Earth, Water, Fire, Wind, and Void. In this tradition *sui* or *mizu*, meaning "water," represents the fluid, formless things in the world. In addition to oceans and waterways, plants are also included under *sui* as they adapt, grow, and change according to the environment, direction of sunlight, and changing seasons. Bodily fluids such as blood are also represented by *sui* as are—interestingly enough—cognitive and emotional tendencies toward adaptation, flexibility, and change.

42. Leanne Shapton, *Swimming Studies* (New York: Penguin, 2012), 188.

43. "Interbeing. What scientific concept would improve everybody's cognitive toolkit?," Edge.org 2001, http://www.edge.org/response-detail/10866.

44. Sally Adee, "Floater," 21 October 2011, The Last Word on Nothing website, http://www.lastwordonnothing.com/2011/10/21/floater.

45. John C. Lilly and Phillip Hensen Bailey Lilly, *The Quiet Center: Isolation and Spirit* (Oakland, CA: Ronin, 2003), 117.

46. Seth Stevenson, "Embracing the Void," 15 May 2013, Slate.com, http://www.slate.com/articles/life/anything_once/2013/05/sensory_deprivation_flotation_tanks_i_floated_naked_in_a_pitch_black_tank.html.

47. Anette Kjellgren, Hanne Buhrkall, and Torsten Norlander, "Preventing sick leave for sufferers of high stress-load and burnout syndrome: a pilot study combining psychotherapy and the flotation tank," *International Journal of Psychology and Psychological Therapy* 11, no. 2 (2011), 297–306.

48. Anette Kjellgren, Hanna Edebol, Tommy Nordén, and Torsten Norlander, "Quality of life with flotation therapy for a person diagnosed with attention deficit disorder, atypical autism, PTSD, anxiety and depression," *Open Journal of Medical Psychology* 2, no. 3 (July 2013), 134–38.

49. Sven Å. Bood, Ulf Sundequist, Anette Kjellgren, Torsten Norlander, Lenneart Nordström, Knut Nordenström, and Gun Nordström, "Eliciting the relaxation response with the help of flotation-REST (restricted environmental stimulation technique) in patients with stress-related ailments," *International Journal of Stress Management* 13, no. 2 (May 2006), 154–75, http://psycnet.apa.org/index.cfm?fa=buy.optionToBuy&id=2006-07100-002.

50. H. Edebol, S. Åke Bood, T. Norlander, "Chronic whiplash-associated disorders and their treatment using flotation-REST (restricted environmental stimulation technique)," *Qualitative Health Research* 18, no. 4 (April 2008), 480–88, http://www.ncbi.nlm.nih.gov/pubmed/18354047.

51. D. G. Forgays and D. K. Forgays, "Creativity enhancement through flotation isolation," *Journal of Environmental Psychology* 12, no. 4 (December 1992), 329–35.

52. Thomas H. Fine and Roderick Borrie, "Flotation REST in Applied Psychophysiology," http://floatforhealth.net/Flotation%20research.htm.

Chapter 5: Blue Mind at Work and Play

1. *2012 Sports, Fitness and Leisure Activities Topline Participation Report,* Sporting Goods Manufacturers Association, 2012 Physical Activity Council. Recreational boating statistic from *2012 Recreational Boating Statistical Abstract*, National Marine Manufacturers Association, 2013. "Of the estimated 232.3 million adults in the U.S. in 2012, 37.8 percent, or 88 million, participated in recreational boating at least once during the year. This is a six percent increase from 2011 and the largest number of U.S. adults participating in boating since NMMA began collecting the data in 1990. Recreational boating participation has steadily increased since 2006."

2. World Health Organization [WHO] Fact Sheet No. 347, "Drowning," October 2012, http://www.who.int/mediacentre/factsheets/fs347/en. And, according to the Centers for Disease Control page "Drowning Risks in Natural Water Settings," "In the U.S. drowning is the second leading cause of unintentional injury death for children ages 1 to 14 years, and the fifth leading cause for people of all ages," http://www.cdc.gov/features/dsdrowningrisks.

3. Centers for Disease Control, *Healthy Swimming Fast Facts*, 2013, http://www.cdc.gov/healthywater/swimming/fast_facts.html.

4. If a 200-pound body is 80 percent water and 15 percent fat, here are the calculations for how much it weighs in water:

$200 \times 20\% = 40$ lbs.
$200 \times 15\% = 30$ lbs.
$40 - 30 = 10$ lbs.

5. Oliver Sacks, "Water Babies," *The New Yorker*, 26 May 1997.

6. See "Healing Waters," by Dr. Bruce E. Becker, *Aquatics International*, June 2007, 27–32.

7. H. Boecker, T. Sprenger, M. E. Spilker, G. Henricksen, M. Koppenhoefer, K. J. Wagner, M. Valet, A. Berthele, and T. R. Tolle, "Runner's High: Opioidergic Mechanisms in the Human Brain," *Cerebral Cortex* 18, no. 11 (2008), 2523–31.

8. B. Draganski and C. Gaser, "Changes in Gray Matter Induced by Training," *Nature* 427, no. 6972 (22 January 2004), 311–12.

9. "In young animals, exercise increases hippocampal neurogenesis and improves learning.... After 1 month, learning was tested in the Morris water maze. Aged runners showed faster acquisition and better retention of the maze than age-matched controls. The decline in neurogenesis in aged mice was reversed to 50% of young control levels by running. Moreover, fine morphology of new neurons did not differ between young and aged runners, indicating that the initial maturation of newborn neurons was not affected by aging." H. van Praag, T. Shubert, C. Zhao, and F. H. Gage, "Exercise Enhances Learning and Hippocampal Neurogenesis in Aged Mice," *Journal of Neuroscience* 25, no. 38 (1 September 2005), 8680–85.

10. C. C. Irwin, R. L. Irwin, N. T. Martin, and S. R. Ross, "Constraints Impacting Minority Swimming Participation: Phase II Qualitative Report," Department of Health and Sports Sciences, University of Memphis, 30 June 2010.

11. D. J. Linden, "Exercise, pleasure and the brain: Understanding the biology of 'runner's high,'" in *The Compass of Pleasure*, excerpted in *Psychology Today*, 21 April 2011, http://www.psychologytoday.com/basics/mating.

12. Zald reports that there are some interesting studies showing that the brains of risk-takers produce and process dopamine differently than those who are less daring. Risk-takers have less of a certain kind of dopamine receptor, which means they need more dopamine to feel the same "hit" of pleasure that an ordinary person gets with less risk.

13. Bridget Reedman, "Scientists Froth on Surf Stoke," 22 November 2011, http://www.theinertia.com/environment/scientists-froth-on-surf-stoke.

14. Source: Surfing Statistics, http://www.statisticbrain.com/surfing.

15. I'll never forget the day world-famous and much beloved "Crocodile Hunter" Steve Irwin died. On September 4, 2006, he had an encounter with an Australian bull ray estimated to weigh about 220 pounds. Irwin was snorkeling in about six feet of water while filming a documentary titled *Ocean's Deadliest* off the coast of Australia with my friend Philippe Cousteau. Irwin was swimming just above the stingray when the animal used the barb on its tail in defense and punctured a hole in his heart. From a distance I provided consoling support to Philippe via e-mail and text messages as he and the crew responded to Steve's injury and death. I can only imagine the tragic turn of the day from euphoric Blue Mind to utter shock and sadness as the world lost a true champion for the ocean and wildlife.

16. For an excellent discussion, see "The Physics of Diving," http://www.scubadiverinfo.com/2_physics.html.

17. Sue Austin's TED talk, "Deep sea diving...in a wheelchair," December 2012, http://www.ted.com/talks/sue_austin_deep_sea_diving_in_a_wheelchair.html.

18. Elizabeth R. Straughan, "Touched by water: the body in scuba diving," *Emotion, Space and Society* 5 (2012), 19–26.

19. Ibid., 22.

20. D. Conradson, "Experiential economies of stillness: The place of retreat in contemporary Britain," in *Therapeutic Landscapes: Geographies of Health,* Allison Williams, ed. (Ashgate: Hampshire, U.K., 2007), 33.

21. W. P. Morgan, "Psychological characteristics of the female diver," in *Women in Diving,* W. P. Fife, ed. (Bethesda, MD: Undersea Medical Society, 1987), 45–54; and W. P. Morgan, "Interaction of anxiety, perceived exertion, and dyspnea in the person-respirator interface," *Medicine and Science in Sports and Exercise* 13 (1981), 73–75.

22. See the Shallow Water Blackout Prevention website, http://shallowwaterblackoutprevention.org.

23. H. Tamaki, K. Kohshi, S. Sajima, J. Takeyama, T. Najamura, H. Ando, T. Ishitake, "Repetitive breath-hold diving causes serious brain injury," *Undersea Hyperbaric Medicine* 37, no. 1 (2010), 8.

24. B. Nevo and S. Breitstein, *Psychological and Behavioral Aspects of Diving* (Flagstaff: Best, 1999), cited in *Psychology of Diving* by Salvatore Capodieci, http://www.psychodive.it/psychology-of-diving13.

25. A. Bonnet, "Régulation émotionnelle et conduites à risques," doctoral dissertation, 2003, Université de Provence, Aix-Marseille I, reported in A. Bonnet, L. Fernandez, A. Piolat, and J. Pedinielli, "Changes in Emotional States Before and After Risk Taking in Scuba Diving," *Journal of Clinical Sport Psychology* 2 (2008), 25–40.

26. *Treatyse on Fysshynge with an Angle*, attributed to Dame Juliana Berners, first published as part of the second edition of *The Boke of St. Albans* in 1496 in England. See also an early article on women fishing by Elizabeth Shaw Oliver, "Angling: One of the Privileges of the Modern Woman," *Country Life in America* 16, no. 2 (June 1909), 171.

27. Diane M. Kuehn, "Elements identified as strongly influencing fishing participation for both males and females were opportunity, perceived ability, and fishing-related customs during childhood..." in "A Discriminant Analysis of Social and Psychological Factors Influencing Fishing Participation," cited in "Proceedings of the 2005 Northeastern Recreation Research Symposium," 10-12 April 2006, General technical report NE, 341, 410–19.

28. For Australian data, see "Identifying the Health and Well-being Benefits of Recreational Fishing," by A. McManus, W. Hunt, J. Storey, and J. White, Centre of Excellence for Science, Seafood and Health, Curtin Health Innovation Research Institute, Curtin University, Perth, Australia, 2011. For U.S. data, see the *2012 Sports, Fitness and Leisure Activities Topline Participation Report,* Sports & Fitness Industry Association.

29. J. Ormsby, "A Review of the Social Motivational and Experiential Char acteristics of Recreational Anglers from Queensland and the Great Barrier Reef Region," Great Barrier Reef Marine Park Authority Research Publication No. 78, February 2004.

30. S. M. Holland and R. B. Ditton, "Fishing Trip Satisfaction: A Typology of Anglers," *North American Journal of Fisheries Management* 12 (1992), 28–32.

31. "Identifying the Health and Well-being Benefits of Recreational Fishing."

32. Ibid., 30–32.

33. Faith Salie, "A baited question: Why do men love fishing?," 15 July 2012, http://www.cbsnews.com/8301-3445_162-57472468/a-baited-question-why-do-men-love-fishing.

34. Brian Fagan, *Beyond the Blue Horizon: How the Earliest Mariners Unlocked the Secrets of the Oceans* (New York: Bloomsbury, 2012), 2.

35. J. Liu, K. Dietz, J. M. DeLoyht, X. Pedre, D. Kelkar, J. Kaur, V. Vialou, M. K. Lobo, D. M. Dietz, E. J. Nestler, J. Dupree, and P. Casaccia, "Impaired adult myelination in the prefrontal cortex of socially isolated mice," *Nature Neuroscience* 15 (2012), 1621–23.

36. "Solo Sailing and Psychological Failure," http://www.associatedglobaltransportservices.com/solo-sailing-psychological-failure.

37. These are extreme cases of isolation, and while a few end in tragedy, it takes a heck of a lot of ocean, time, and stress to send someone completely

over the edge. Time spent at sea for the other 99.999 percent is a healing and enjoyable experience.

38. From 2000 to 2010, commercial fishers averaged 124 deaths per 100,000 workers, compared with 4 per 100,000 among all U.S. workers. U.S. Department of Labor, Bureau of Labor Statistics, "Injuries, Illnesses and Fatalities: Census of Fatal Occupational Injuries," current and revised data, Washington, D.C. (2012).

39. M. Murray, "The Use of Narrative Theory in Understanding and Preventing Accidents in the Fishing Industry," in "International Fishing Industry Safety and Health Conference," 24 October 2000, 245.

40. D. Sneed, "Morro Bay fishing is back from the depths," *The Tribune*, 3 July 2010, http://www.sanluisobispo.com/2010/07/03/1203763/morro-bay-fishing-is-back-from.html#storylink=cpy.

Chapter 6: Red Mind, Gray Mind, Blue Mind: The Health Benefits of Water

1. W. Zhong, H. Maradit-Kremers, J. L. St. Sauver, B. P. Yawn, J. O. Ebbert, V. L. Roger, D. J. Jacobson, M. E. McGree, S. M. Brue, W. A. Rocca, "Age and sex patterns of drug prescribing in a defined American population," *Mayo Clinic Proceedings* 88, no. 7 (July 2013), 697–707, http://www.mayoclinicproceedings.org/article/S0025-6196%2813%2900357-1/abstract. The researchers studied prescription records for Olmstead County, Minnesota, in 2009 and, according to Jennifer St. Sauver, the results were comparable to those found elsewhere in the United States. See "Nearly 7 in 10 Americans are on prescription drugs," *Science Daily*, 6 June 2013, http://www.sciencedaily.com/releases/2013/06/130619132352.htm.

2. I. Kirsch, B. J. Deacon, T. B. Huedo-Medina, A. Scorboria, T. J. Moore, and B. T. Johnson, "Initial severity and antidepressant benefits: A meta-analysis of data submitted to the Food and Drug Administration," *PLoS Medicine*, 26 February 2008, doi: 10.1371/journal.pmed.0050045, http://www.plosmedicine.org/article/info:doi/10.1371/journal.pmed.0050045.

3. J. J. Radley, A. B. Rocher, M. Miller, W. G. Janssen, C. Liston, P. R. Hopf, B. S. McEwen, and J. H. Morrison, "Repeated stress induces dendritic spine loss in the rat medial prefrontal cortex," *Cerebral Cortex* 16, no. 3 (March 2006), 313–20, http://www.ncbi.nlm.nih.gov/pubmed/15901656.

4. B. S. McEwen and S. Chattarji, "Molecular mechanisms of neuroplasticity and pharmacological implications: The example of tianeptine," *European Neuropsychopharmacology* 14, suppl. 5 (December 2004), S497–502, http://www.ncbi.nlm.nih.gov/pubmed/15550348.

5. R. M. Sapolsky, "Gluocorticoids and hippocampal atrophy in neuropsychiatric disorders," *Archives of General Psychiatry* 57, no.10 (October 2000), 925–35.

6. A. Keller, K. Litzelman, et al., "Does the Perception that Stress Affects Health Matter?," *Health Psychology* 31, no. 5 (September 2012), 677–84.

7. My colleague Julia Townsend was recently prescribed surfing by her allergist, with excellent results. She wrote:

"I've had asthma and allergies my entire life. My biggest allergen is an indoor irritant, dust. When I moved from cold and frosty New England to the Monterey Bay Area six and a half years ago I anticipated that my allergies would be less intense. I quickly learned that Monterey is one of the worst places in the country for dust allergies. Why? My understanding is that dust grows in dark, damp, cold areas. Cue the fog that accompanies the local weather year-round. A couple of years ago, while seeking help with curtailing my allergies I was lucky enough to be referred to Dr. Jeffrey Lehr. Among other strategies, he mentioned that I should surf more. His reasoning as I recall was that the ocean can give allergy sufferers a break from land and household based allergens. For those with rhinitis saltwater delivers the additional benefit of soothing nasal passages. I have remembered Dr. Lehr's advice and used it in conjunction with traditional medicine and household upkeep. I have also had primary care physicians that have recommended surfing as a tool for stress management."

This idea is spreading fast. The Golden Gate National Parks Conservancy has a collaborative program with medical professionals called Park Prescriptions, http://www.parksconservancy.org/assets/programs/igg/pdfs/park-prescriptions-2010.pdf.

8. M. White, A. Smith, et al., "Blue space: the importance of water for preference, affect, and restorativeness ratings of natural and built scenes," *Journal of Environmental Psychology* 30 (2010), 482–93.

9. Risk-seekers usually score high for novelty-seeking. Dr. David Zald, who studies dopamine and addiction, has shown that many risk-takers actually have fewer inhibitors for dopamine in their brains, which means they receive a greater "rush" from new or risky behaviors. See chapters 4 and 6.

10. According to the Centers for Disease Control, in 2009 the top ten causes of death in the United States, in order, were heart disease, cancer, respiratory disease, stroke, accidents, Alzheimer's disease, diabetes, influenza and pneumonia, kidney disease, and suicide. All of those diseases can be a result of, or exacerbated by, chronic stress. See "Psychological Stress and Disease" by Sheldon Cohen, Ph.D., Denise Janicki-Deverts, Ph.D., and Gregory E. Miller, Ph.D., *Journal of the American Medical Association* 298, no. 14 (10 October 2007), 1685–87.

11. Studies cited in Rick Hanson, *Buddha's Brain: The Practical Neuroscience of Happiness, Love and Wisdom* (Oakland, CA: New Harbinger, 2009), 56–58.

12. Amy Arnsten, Carolyn M. Mazure, and Rajita Sinha, "This is Your Brain in Meltdown," *Scientific American 306* (April 2012), 48–53, published online: 20 March 2012.

13. Alex Soojung-Kim Pang, *The Distraction Addiction* (New York and Boston: Little, Brown and Company, 2013), 10.

14. Ibid., 11.

15. Daniel Goleman, *Focus: The Hidden Driver of Excellence* (New York: HarperCollins, 2013), 203.

16. Kate Parkinson-Morgan, "Anthropology professor Monica Smith investigates multitasking as ancient ability in new book," *Daily Bruin,* UCLA, 7 January 2011.

17. https://www.sciencenews.org/article/impactful-distraction.

18. Pang, *The Distraction Addiction*, 59.

19. Peter Bregman, *18 Minutes: Find Your Focus, Master Distraction, and Get the Right Thiings Done* (New York: Business Plus, 2011), 122–23.

20. https://www.sciencenews.org/article/impactful-distraction.

21. http://lindastone.net/qa.

22. Pang, *The Distraction Addiction*, 64.

23. Doidge, *The Brain That Changes Itself,* 68 (see chap. 2, n. 26)

24. Eyal Ophir, Clifford Nass, and Anthony D. Wagner, "Cognitive control in media multitaskers," *Proceedings of the National Academy of Sciences* 106, no. 37 (15 September 2009), 15583–87, http://www.pnas.org/content/early/2009/08/21/0903620106.abstract.

25. Bregman, *18 Minutes*, 220–21.

26. Douglas T. Kenrick, "Subselves and the Modular Mind," in *This Will Make You Smarter*, 123–34.

27. For a discussion of the psychological benefits experienced by many extreme athletes, see "The Psychology of Extreme Sports: Addicts, not Loonies," by Joachim Vogt Isaksen, http://www.popularsocialscience.com/2012/11/05/the-psychology-of-extreme-sports-addicts-not-loonies.

28. Laura Parker Roerden, "Your Mind on Blue and a 'Lucky' Karina Dress Giveaway," Ocean Matters blog, 24 May 2013, http://www.oceanmatters.org/category/staff.

29. S. Kaplan, "The restorative benefits of nature: Toward an integrative framework," *Journal of Environmental Psychology* 15 (1995), 169–82.

30. Quoted in Eric Jaffe, "This Side of Paradise: Discovering Why the Human Mind Needs Nature," Association for Psychological Science *Observer*

23, no. 5 (May/June 2010), http://www.psychologicalscience.org/index.php/publications/observer/2010/may-june-10/this-side-of-paradise.html.

31. Marc G. Berman, John Jonides, and Stephen Kaplan, "The Cognitive Benefits of Interacting with Nature," *Psychological Science* 19, no. 2 (December 2008), 1207–12.

32. See "The Brain in the City," by Richard Coyne, http://richardcoyne.com/2013/03/09/the-brain-in-the-city. This article was based upon the study "The urban brain: analysing outdoor physical activity with mobile EEG," by Peter Aspinall, Panagiotis Mavros, Richard Coyne, and Jenny Roe, *British Journal of Sports Medicine,* 6 March 2013, http://bjsm.bmj.com/content/early/2013/03/05/bjsports-2012-091877.abstract.

33. In 1980 Calgon produced a TV ad that has stuck with me. An actress begins by listing the stresses of her life and then escapes to the serenity of her bath. Woman: "The traffic, the boss, the baby, the dog, that does it! Calgon, take me away!" Announcer: "Lose your cares in the luxury of a Calgon bath… as it lifts your spirits," http://www.youtube.com/watch?v=fJsnR-KDbFc.

34. Hippocrates, *On Airs, Waters, and Places,* part 7, trans. Francis Adams. Made available through the Internet Classics archive, © 1994–2000, Daniel C. Stevenson, Web Atomics, http://classics.mit.edu//Hippocrates/airwatpl.html.

35. M. Toda, K. Morimoto, S. Nagasawa, and K. Kitamura, "Change in salivary physiological stress markers by spa bathing," *Biomedical Research* 27, no. 11 (2006), 11–14.

36. K. Mizuno, M. Tanaka, K. Tajima, N. Okada, K. Rokushima, and Y. Watanabe, "Effects of mild-stream bathing on recovery from mental fatigue," *Medical Science Monitor* 16, no. 1 (January 2010), 8–14, http://www.ncbi.nlm.nih.gov/pubmed/20037494.

37. B. A. Levine, "Use of hydrotherapy in reduction of anxiety," *Psychological Reports* 55, no. 2 (October 1984), 526.

38. J. S. Raglin and W. P. Morgan, "Influence of vigorous exercise on mood states," *Behavior Therapist* 8 (1985), 179–83, cited in J. D. Kreme and D. Scully, *Psychology in Sport* (Florence, KY: Routledge, 1994), 176.

39. See R. D. Benfield, T. Hortobágyi, C. J. Tanner, M. Swanson, M. M. Heitkemper, and E. R. Newton, "The effects of hydrotherapy on anxiety, pain, neuroendocrine responses, and contraction dynamics during labor," *Biologic Research for Nursing* 12, no. 1 (July 2010), 28–36, http://brn.sagepub.com/content/12/1/28.short. Showers also have proven effective in decreasing tension and anxiety, and increasing levels of relaxation; see M. A. Stark, "Therapeutic showering in labor," *Clinical Nursing Research* 22, no. 3 (August 2013), 359–74, http://cnr.sagepub.com/content/22/3/359.

40. J. Hall, S. M. Skevington, P. J. Maddison, and K. Chapman, "A randomized and controlled trial of hydrotherapy in rheumatoid arthritis," *Arthritis Care and Research* 9, no. 3 (June 1996), 206–15, http://www.ncbi.nlm.nih.gov/pubmed/8971230.

41. See Y. Saeki, "The effect of foot-bath with or without the essential oil of lavender on the autonomic nervous system: a randomized trial," *Complementary Therapies in Medicine* 8, no. 1 (March 2000), 2–7, http://www.ncbi.nlm.nih.gov/pubmed/10812753.

42. V. Neelon and M. Champagne, "Managing cognitive impairment: the current bases for practice," in *Key Aspects of Eldercare: Managing Falls, Incontinence and Cognitive Impairment*, S. Funk, E. Tournquist, and M. Champagne, eds. (New York: Springer, 1992), 122–31.

43. Summary in C. J. Edmonds, R. Crombie, and M. R. Gardner, "Subjective thirst moderates changes in speed of responding associated with water consumption," *Frontiers in Human Neuroscience* 7, art. 363 (2013), published online 16 July 2013, http://www.ncbi.nlm.nih.gov/pmc/articles/PMC3712897.

44. M-M. G. Wilson and J. E. Morley, "Impaired cognitive function and mental performance in mild dehydration," *European Journal of Clinical Nutrition* 57, suppl. 2 (2003), S24–29, http://www.nature.com/ejcn/journal/v57/n2s/full/1601898a.html.

45. M-M. G. Wilson and J. E. Morley, citing J. L. Warren, W. E. Bacon, T. Harris, A. M. McBean, D. J. Foley, and C. Phillips, "The burden and outcomes associated with dehydration among US elderly," *American Journal of Public Health* 84 (1994), 1265–69; and D. K. Miller, H. M. Perry, and J. E. Morley, "Relationship of dehydration and chronic renal insufficiency with function and cognitive status in older US blacks," in *Hydration and Aging: Facts, Research, and Intervention in Geriatric Series*, B. Vellas, J. L. Albarede, and P. J. Garry, eds. (New York: Serdi and Springer, 1998), 149–59.

46. C. J. Edmonds, R. Crombie, and M. R. Gardner, citing Y. Bar-David, J. Urkin, and E. Kozminsky, "The effect of voluntary dehydration on cognitive functions of elementary school children," *Acta Paediatrica* 94, no. 11 (November 2005), 1667–73, http://www.ncbi.nlm.nih.gov/pubmed/16303708; R. Fadda, G. Rappinett, D. Grathwohl, M. Parisi, R. Fanari, C.M. Caio, et al., "Effects of drinking supplementary water at school on cognitive performance in children," *Appetite* 59, no. 3 (2012), 730–37, http://www.ncbi.nlm.nih.gov/pubmed/22841529; F. Bonnet, E. Lepicard, L. Cathrin, C. Letellier, F. Constant, N. Hawili, et al., "French children start their school day with a hydration deficit," *Annals of Nutrition and Metabolism* 60, no. 4 (2012), 257–63, http://www.ncbi.nlm.nih.gov/pubmed/22677981; G. Fried-

lander, "Hydration Status of Children in the US and Europe, Optimal Hydration: New Insights," Academy of Nutrition and Dietetics Food and Nutrition Conference and Expo (Philadelphia: Nestlé Nutrition Institute, 2012), available online at: http://www.nestlenutrition-institute.org/Events/All_Events/Documents/ADA%202012/NNI_FNCE_PpP.pdf; and J. Stookey, B. Brass, A. Holliday, A.I. Arieff, "What is the cell hydration status of healthy children in the USA? Preliminary data on urine osmolality and water intake," *Public Health Nutrition* 15, no. 11 (2012), 2148–56, http://www.ncbi.nlm.nih.gov/pubmed/22281298.

47. C. J. Edmonds, R. Crombie, and M. R. Gardner, citing C. J. Edmonds and D. Burford, "Should children drink more water? The effects of drinking water on cognition in children," *Appetite* 52, no. 3 (2009), 776–79, http://www.ncbi.nlm.nih.gov/pubmed/19501780; C. J. Edmonds and B. Jeffes, "Does having a drink help you think? 6–7 year old children show improvements in cognitive performance from baseline to test after having a drink of water," *Appetite* 53, no. 3 (2009), 469–72, http://www.ncbi.nlm.nih.gov/pubmed/19835921; and P. Booth, B. G. Taylor, and C. J. Edmonds, "Water supplementation improves visual attention and fine motor skills in schoolchildren," *Education and Health* 30, no. 3 (2012), 75–79, http://sheu.org.uk/x/eh303pb.pdf.

48. As of 2013, according to statistics compiled by *Beverage Digest*, U.S. adults consumed 58 gallons of water per year, which equals around 20 ounces per day. They also drank 44 gallons of soda per year—down from 54 gallons in 1998. See J. Hamblin, "How much water do people drink?" *The Atlantic* online, 12 March 2013, http://www.theatlantic.com/health/archive/2013/03/how-much-water-do-people-drink/273936.

49. Quoted in H. Mount, "The summer holidays—and where we do like to be," *Telegraph* online, 17 August 2012, http://www.telegraph.co.uk/lifestyle/9482952/The-summer-holidays-and-where-we-do-like-to-be.html.

50. R. Knowles, "George III in Weymouth," 27 July 2012, http://www.regencyhistory.net/2012/07/george-iii-in-weymouth.html.

51. M. H. Depledge and W. J. Bird, "The Blue Gym: Health and wellbeing from our coasts," *Marine Pollution Bulletin* 58, no. 7 (July 2009), 947–48.

52. "Professor Mike Depledge Blue Gym.mov," uploaded 1 December 2009, https://www.youtube.com/watch?v=Jq44KhBSgQA.

53. B. M. Wheeler, M. White, W. Stahl-Timmins, M. H. Depledge, "Does living by the coast improve health and wellbeing?" *Health & Place* 18, no. 5 (September 2012), 1198–1201, http://www.sciencedirect.com/science/article/pii/S1353829212001220.

54. K. J. Ashbullby, S. Pahl, P. Webley, and M. P. White, "The beach as a setting for families' health promotion: A qualititative study with parents and

children living in coastal regions in Southwest England," *Health & Place* 23 (September 2013), 138–47, http://www.sciencedirect.com/science/article/pii/S1353829213000877.

55. A. Bauman, B. Smith, L. Stoker, B. Bellew, and M. Booth, "Geographical influences upon physical activity participation: evidence of a 'coastal effect,'" *Australia and New Zealand Journal of Public Health* 23, no. 3 (June 1999), 322–24.

56. Wheeler et al., "Does living by the coast," 1200.

57. C. J. Thompson, K. Boddy, K. Stein, R. Whear, J. Barton, and M. H. Depledge, "Does participating in physical activity in outdoor natural environments have a greater effect on physical and mental wellbeing than physical activity indoors? A systematic review," *Environmental Science & Technology* 45, no. 5 (March 2011), 1761–72, http://www.ncbi.nlm.nih.gov/pubmed/21291246.

58. J. Barton and J. Pretty, "What is the best dose of nature and green exercise for improving mental health? A multi-study analysis," *Environmental Science & Technology* 15, no. 44 (May 2010), 3947–55, http://www.ncbi.nlm.nih.gov/pubmed/20337470.

59. *Heroes on the Water* video, https://www.youtube.com/watch?v=iZQFKPaVNMQ.

60. Hurricane Sandy created just such a problem for Captain Joel Fogel, who wrote that he was devastated by the effects of the storm on the Barrier Beach Islands in New Jersey. "It's a year later and I still get a little sick when I watch the ocean," he said. And following the 2010 BP oil spill in the Gulf of Mexico, thousands of people in Louisiana, Mississippi, Alabama, Texas, and Florida reported symptoms of post-traumatic stress.

61. T. Tanielian and L. H. Jaycox, eds., *Invisible Wounds of War: Psychological and Cognitive Injuries, Their Consequences, and Services to Assist Recovery* (Santa Monica: RAND Corporation, 2008), 12.

62. In sleep, explains Michio Kaku, "when the dorsolateral prefrontal cortex is shut down, we can't count on the rational, planning center of the brain. Instead, we drift aimlessly in our dreams, with the visual center giving us images without rational control. The orbitofrontal cortex, or the fact-checker, is also inactive. Hence dreams are allowed to blissfully evolve without any constraints from the laws of physics or common sense. And the temporoparietal lobe, which helps coordinate our sense of where we are located using signals from our eyes and inner ear, is also shut down, which may explain our out-of-body experiences while we dream." Michio Kaku, *The Future of the Mind: The Scientific Quest to Understand, Enhance, and Empower the Mind* (New York: Doubleday, 2014), 173.

63. H. Frumkin, "Beyond toxicity: Human health and the natural environment," *American Journal of Preventive Medicine* 20, no. 3 (2001), 234–40.

64. "Rivers of Recovery: Results," http://www.riversofrecovery.org/what-we-do/medical-research/results.

65. Jaimal Yogis, *The Fear Project: What Our Most Primal Emotion Taught Me About Survival, Success, Surfing . . . and Love* (New York: Rodale, 2012), 179.

66. "Darryl 'Flea' Virostko," http://fleahab.net/about-fleahab.

67. E. Jeffries, "The pleasure principle," *Nature Climate Change* 3, no. 9 (September 2013), 776–77.

68. See Winifred Gallagher, *The Power of Place,* 137–38 (see chap. 3, n. 14).

69. D. S. Vonder Hulls, L. K. Walker, and M. Powell, "Clinicians' perceptions of the benefits of aquatic therapy for young children with autism: A preliminary study," *Physical & Occupational Therapy in Pediatrics* 26, nos.1–2 (2006), 13–22.

70. C. Pan, "Effects of water exercise swimming program on aquatic skills and social behaviors in children with autism spectrum disorders," *Autism* 14, no. 1 (January 2010), 9–28.

71. S. D. James, "Surfing turns autistic kids into rock stars," ABC News via *Good Morning America,* 20 August 2013, http://abcnews.go.com/Health/surfing-transforms-autistic-kids-rock-stars/story?id=20004325.

72. Kara Collins, on the Surfers for Autism website, http://surfersforautism.org/?page_id=19.

73. L. Jake, "Autism and the role of aquatic therapy in recreational therapy treatment services," 1 September 2003, http://www.recreationtherapy.com/articles/autismandquatictherapy.htm.

74. "Autistic children crave being in water," WestBend Culture of Safety, 29 April 2011, http://www.cultureofsafety.com/2011/04/autistic-children-crave-being-in-water.

75. Naoki Higashida, *The Reason I Jump* (New York: Random House, 2007; trans. 2013), 71–72.

76. http://www.patagonia.com.au/journal/2012/changing-lives.

77. J. Soboroff, interviewer, "Ocean a comfort zone for children with autism," *Huffpost Live,* originally aired 17 November 2012, http://live.huffingtonpost.com/r/segment/autism-in-the-water/509bffb72b8c2a65e1000492.

Chapter 7: Blue Unified: Connection and Water

1. Michael J. Fox, *Lucky Man: A Memoir* (New York: Hyperion, 2002), 242–43.

2. http://www.today.com/entertainment/michael-j-fox-i-never-really-went-anywhere-4B11248002.

3. William James, "The Varieties of Religious Experience: A Study in Human Nature," in *William James: Writings 1902–1910* (New York: Library of America, 1988), 349.

4. http://bigstory.ap.org/article/mindfulness-grows-popularity-and-profits.

5. David Gelles, "The Mind Business," *Financial Times,* 24 August 2012.

6. Clint Eastwood's 1971 directorial debut, *Play Misty for Me,* features a romantic montage of the lead actors (Eastwood and Donna Mills) strolling by the Carmel, California, seaside, backed by Roberta Flack's recording of the Ewan MacColl torch song "The First Time Ever I Saw Your Face." The bucolic scene sharply contrasts with the rest of this psychological thriller, whose tagline was "The scream you hear may be your own," http://www.youtube.com/watch?v=ypSPbIAApuQ.

7. "Management in the Age of Complexity," *Harvard Business Review,* March 2014.

8. Pang, *The Distraction Addiction,* 83 (see chap. 6, n. 13).

9. Goleman, *Focus,* 17 (see chap. 6, n. 15).

10. S. McMains and S. Kastner, "Interactions of top-down and bottom-up mechanisms in human visual cortex," *Journal of Neuroscience* 31, no. 2 (12 January 2011), 587–97, http://www.ncbi.nlm.nih.gov/pubmed/21228167.

11. N. H. Mackworth, "The breakdown of vigilance during prolonged visual search." *Quarterly Journal of Experimental Psychology* 1 (1948), 6–21.

12. In a *Fast Company* blog post, Laura Vanderkam gives an example of one company's z-mail policy (http://www.fastcompany.com/3019655/how-to-be-a-success-at-everything/should-your-company-practice-zmail-the-case-for-inbox-curf). It would be excruciatingly hard to come up with many more such American or British firms. In France, a recent labor deal declared that workers have no obligation to look at their work-related e-mail, or answer or check their phones, outside of work hours; this includes employees working for some of the same social media and web-based companies that contribute to our Red Mind distraction, http://www.nbcnews.com/business/careers/pardonnez-moi-boss-i-cant-answer-phone-now-im-chez-n77106. In Germany, companies such as VW, Puma, and BMW have all established policies to restrict out-of-office e-mailing, http://www.telegraph.co.uk/news/worldnews/europe/germany/10276815/Out-of-hours-working-banned-by-German-labour-ministry.html?utm_source=nextdraft&utm_medium=email.

13. http://www.fastcodesign.com/3020896/asides/addicted-to-your-phone-its-your-fault-wired-says.

14. Jeffrey Kluger, "Accessing the creative spark," *Time,* 9 May 2013, http://business.time.com/2013/05/09/assessing-the-creative-spark.

15. S. Liu, H. M. Chow, Y. Xu, M. G. Erkkinen, K. E. Swett, M. W. Eagle, D. A. Rizik-Baer, and A. R. Braun, "Neural correlates of lyrical improvisation: an fMRI study of freestyle rap," *Scientific Reports* 2, art. 834 (15 November 2012), http://www.nature.com/srep/2012/121115/srep00834/full/srep00834.html.

16. Ibid.

17. D. W. Winnicott, "Transitional objects and transitional phenomena: a study of the first not-me possession," *International Journal of Psychoanalysis* 34, no. 2 (1953), 89–97, http://nonoedipal.files.wordpress.com/2009/09/transitional-objects-and-transitional-phenomenae28094a-study-of-the-first-not-me-possession.pdf.

18. Alexandra Enders, "The importance of place: where writers write and why," *Poets & Writers,* March/April 2008, http://www.pw.org/content/importance_place_where_writers_write_and_why_0?cmnt_all=1.

19. Gallagher, *The Power of Place,* 133 (see chap. 3, n. 14).

20. "The work habits of highly successful writers," *Postscripts,* 23 May 2006, http://notorc.blogspot.com/2006/05/work-habits-of-highly-successful_23.html.

21. Alexandra Alter, "How to write a great novel," *The Wall Street Journal,* 13 November 2009, http://online.wsj.com/article/SB10001424052748703740004574513463106012106.html.

22. David Wallace-Wells, "A brain with a heart," *New York,* 4 November 2012, http://nymag.com/news/features/oliver-sacks-2012-11.

23. Oliver Sacks, "Water babies: why I love to swim," *The New Yorker,* 26 May 1997, http://archives.newyorker.com/?iid=15566&crd=0&searchKey=Water%20Babies#folio=044.

24. http://www.bbc.co.uk/programmes/b03ybpf8.

25. Ariane Conrad, "Water, water everywhere: Ran Ortner's love affair with the sea," *The Sun,* no. 438 (June 2012), http://thesunmagazine.org/issues/438/water_water_everywhere.

26. B. R. Conway and A. Rehding, "Neuroaesthetics and the trouble with beauty," *PLoS Biology* 11, no. 3 (19 March 2013), e1001504. doi:10.1371/journal.pbio.1001504, http://www.plosbiology.org/article/info%3Adoi%2F10.1371%2Fjournal.pbio.1001504.

27. Ibid.

28. Leonardo da Vinci was fascinated by water, yet his fascination focused more on the physical and ecological aspects of water than the cognitive and emotional aspects: "Water is sometimes sharp and sometimes strong, sometimes acid and sometimes bitter, sometimes sweet and sometimes thick or thin, sometimes it is seen bringing hurt or pestilence, sometimes health-giving, sometimes poisonous. It suffers change into as many

natures as are the different places through which it passes. And as the mirror changes with the color of its subject, so it alters with the nature of the place, becoming noisome, laxative, astringent, sulfurous, salty, incarnadined, mournful, raging, angry, red, yellow, green, black, blue, greasy, fat or slim. Sometimes it starts a conflagration, sometimes it extinguishes one; is warm and is cold, carries away or sets down, hollows out or builds up, tears or establishes, fills or empties, raises itself or burrows down, speeds or is still; is the cause at times of life or death, or increase or privation, nourishes at times and at others does the contrary; at times has a tang, at times is without savor, sometimes submerging the valleys with great floods. In time and with water, everything changes."

29. While many believe that "paying too much attention to those artsy-fartsy, touchy-feely elements will eventually dumb us down and screw us up..." writes Daniel Pink in *A Whole New Mind: Why Right-Brainers Will Rule the Future* (New York: Penguin, 2006), "... the future belongs to a very different kind of person with a very different kind of mind—creators and empathizers, pattern recognizers, and meaning makers. These people—artists, inventors, designers, storytellers, caregivers, consolers, big picture thinkers—will now reap society's richest rewards and share its greatest joys" (17, 1).

30. My brave brother-in-law Jon Imber continued to paint by swaying his entire body and eventually by attaching a paintbrush to his head. Diagnosed with ALS, or Lou Gehrig's disease, in the fall of 2012, he soon lost the use of his hands and arms. "To this artist, it's evident that to paint is to live," http://www.pressherald.com/news/Jon_Imber_renowned_New_England_painter_dies_html.

31. P. Stokowski, "Symbolic aspects of water," in *Water and People: Challenges at the Interface of Symbolic and Utilitarian Values*, S. F. McCool, R. J. Clark, and G. H. Stankey, eds. (Washington, D.C.: United States Department of Agriculture, Forest Service, January 2008), 32.

32. Ivan Illich, *H_2O and the Waters of Forgetfulness* (London: Marion Boyars, 2000), 24–25.

33. D. Hofstadter and E. Sander, "Analogy: The vital talent that fuels our minds," *New Scientist*, 2915 (9 May 2013), http://www.newscientist.com/article/mg21829150.400-analogy-the-vital-talent-that-fuels-our-minds.html.

34. Ibid.

35. David A. Havas and James Matheson, "The Functional Role of the Periphery in Emotional Language Comprehension," *Frontiers in Psychology*, 27 May 2013, doi: 10.3389/fpsyg.2013.00294.

36. J. Ackerman, C. Nocera, and J. Bargh, "Incidental Haptic Sensations Influence Social Judgments and Decisions," *Science* 238 (25 June 2010), 1712.

37. See D. Western, P. S. Blagov, K. Harenski, C. Kilts, and S. Hamann, "Neural bases of motivated reasoning: an fMRI study of emotional constraints on partisal political judgment in the 2004 U.S. presidential election," *Journal of Cognitive Neuroscience* 18, no. 11 (2006), 1947–58, http://psychsystems.net/lab/06_Westen_fmri.pdf.

38. Goleman, *Focus*, 43 (see chap. 6, n. 15).

39. M. Slepian and N. Ambady, "Fluid Movement and Creativity," *Journal of Experimental Psychology* 141, no. 4 (November 2012), 625–29.

40. Ibid.

41. Keith J. Holyoak and Paul Thagard, *Mental Leaps: Analogy in Creative Thought* (Boston: MIT Press, 1995), 12.

42. D. Franklin, "How hospital gardens help patients heal" (published in print as "Nature that Nurtures"), *Scientific American* 306 (March 2012), 24–25, http://www.scientificamerican.com/article.cfm?id=nature-that-nurtures.

43. S. Sherman, J. Varni, R. Ulrich, V. Malcarne, "Post-occupancy evaluation of healing gardens in a pediatric cancer center," *Landscape and Urban Planning* 73, no. 2 (October 2005), 167–83.

44. R. S. Ulrich, O. Lunden, and J. L. Eltinge, "Effects of exposure to nature and abstract pictures on patients recovering from heart surgery," *Psychophysiology* 30, suppl. 1 (1993), 7, cited in S. Mitrione, "Therapeutic responses to natural environments: Using gardens to improve health care," *Minnesota Medicine* 91, no. 3 (March 2008), 31–34, http://www.minnesotamedicine.com/PastIssues/PastIssues2008/March2008/ClinicalMitrioneMarch2008/tabid/2488/Default.aspx.

45. One ongoing study is tracking the responses of people who are shown scenes of blue (water) environments while cycling on stationary bicycles indoors. Mat White reports that there is "a lot less activity in the brain when the sea is shown... which tells us that it's probably less stressful and more familiar to the core human being."

46. D. Cracknell, M. P. White, S. Pahl, W. J. Nichols, and M. H. Depledge, "Sub-aquatic biodiversity and psychological well-being: a preliminary examination of dose-response effects in an aquarium setting," *Journal of Environmental Psychology*, in press.

47. A. Katcher, H. Segal, and A. Beck, "Comparison of contemplation and hypnosis for the reduction of anxiety and discomfort during dental surgery," *American Journal of Clinical Hypnosis* 27, no. 1 (1984), 14–21, http://www.tandfonline.com/doi/abs/10.1080/00029157.1984.10402583#preview.

48. Bryan C. Pijanowski, Luis J. Villanueva-Rivera, Sarah L. Dumyahn, Almo Farina, Bernie L. Krause, Brian M. Napoletano, Stuart H. Gage, and Nadia Pieretti, "Soundscape Ecology: The Science of Sound in the Landscape," *BioScience* 61, no. 3 (March 2011), 203–16, http://www.jstor.org/stable/10.1525/bio.2011.61.3.6.

49. Lisa Goines and Louis Hagler, "Noise Pollution: A Modern Plague," *Southern Medical Journal,* 100 (March 2007), 287–94.

50. E. van Kempen and W. Babisch, "The quantitative relationship between road traffic noise and hypertension: a meta-analysis," *Journal of Hypertension* 30, no. 6 (June 2012), 1075–86; M. Sorensen, Z. J. Andersen, R. B. Nordsborg, S. S. Jensen, K. G. Lillelund, R. Beelen, E. B. Schmidt, A. Tjonneland, K. Overvad, and O. Raaschou-Nielsen, "Road traffic noise and incident myocardial infarction: a prospective cohort study, *PLoS One* 7, no. 6 (20 June 2012), http://www.plosone.org/article/info%3Adoi%2F10.1371%2Fjournal.pone.0039283; D. Prasher, "Is there evidence that environmental noise is immunotoxic?," *Noise Health* 11, no. 44 (July-September 2009), 151–55.

51. W. Passchier-Vermeer and W. F. Passchier, "Noise exposure and public health." *Environmental Health Perspectives* 108, supp. 1 (March 2000), 123–31.

52. While there are many studies on the effects of music on mood, relaxation, and concentration, the number of studies utilizing nature sounds is much smaller. One is J. J. Alvarsson, S. Wiens, and M. E. Nilsson, "Stress recovery during exposure to nature sound and environmental noise," *International Journal of Environmental Research and Public Health* 7, no. 3 (March 2010), 1036–46.

53. The sounds of oceans, rain, wind, and so on usually have more energy at low frequencies than at high (Voss and Clarke, 1975), and feature slow temporal modulations (Attias & Schreiner, 1997; Singh & Theunissen, 2003), rather than prominent modulations in the roughness range.

54. Chiba University researcher Dr. Yoshifumi Miyazaki and colleagues demonstrated that creek sounds can induce changes in blood flow in the brain indicative of a relaxed state, opposite of what is encountered during mental and cognitive stress and exhaustion. Y. Tsunetsugu, B.-J. Park, and Y. Miyazaki, "Trends in research related to 'Shinrin-yoku' (taking in the forest atmosphere or forest bathing) in Japan," *Environmental Health and Preventive Medicine* 15, no. 1 (January 2010), 27–37. Also described in E. M. Selhub and A. C. Logan, *Your Brain On Nature: The Science of Nature's Influence on Your Health, Happiness and Vitality* (Mississauga, Ontario: Wiley, 2012), 97.

55. A 1992 study in Huntsville, Alabama, had postoperative coronary artery bypass graft patients listen to ocean sounds at night in the recovery ward. They reported better sleep depth, awakening, return to sleep, quality

of sleep, and total sleep scores. J. W. Williamson, "The effects of ocean sounds on sleep after coronary artery bypass graft surgery," *American Journal of Critical Care* 1, no. 1 (July 1992), 91–97, http://www.ncbi.nlm.nih.gov/pubmed/1307884.

56. A. N. Abd El Aziz, K. Jahangir, Y. Kobayashi, F. Norliyana, and Jamil A. Amad, "Evaluation of the Effect of Preoperative Natural Water Fountain Melody on Teenagers' Behavior—Preliminary Study," *American Journal of Sociological Research* 2, no. 4 (2012), 78–81, http://article.sapub.org/10.5923.j.sociology.20120204.04.html.

57. M. D. Hunter, S. B. Eickhoff, R. J. Pheasant, M. J. Douglas, G. R. Watts, T. F. D. Farrow, D. Hyland, J. Kang, I. D. Wilkinson, K. V. Horoshenkov, and P. W. R. Woodruff, "The state of tranquility: Subjective perception is shaped by contextual modulation of auditory connectivity," *Neuroimage* 53 (2010), 611.

58. L. S. Berk and B. Bittman, "A video presentation of music, nature's imagery and positive affirmations as a combined eustress paradigm modulates neuroendocrine hormones," *Annals of Behavioral Medicine* 19, suppl. (1997), 174.

59. Jian Kang, *Urban Sound Environment* (New York: Taylor & Francis, 2007).

60. Top-selling sound machines feature ocean, rain, waterfall, heartbeat, and rainforest settings.

61. Just as the smiley face emoticon :-) can now elicit a neural response similar to that of a human face, so might those small fishbowls with faux pirate treasure and ceramic sea turtles suggest their much larger real-world analogue. Douglas Main, "Human Brains Now Understand Smiley Emoticon Like A Real Face," *Scientific American*, 10 February 2014, http://www.popsci.com/article/science/human-brains-now-understand-smiley-emoticon-real-face.

62. Lumosity website, http://www.lumosity.com/about, accessed 22 February 2014.

63. Posit Science website, http://www.positscience.com/, accessed 22 February 2014.

64. At the 2014 Digital Kids Conference, the media research firm the Michael Cohen Group presented results of a nationwide survey that polled 350 parents about play habits of their children twelve and younger. Touchscreens ranked highest, with 60 percent of parents claiming their child uses a touchscreen "often" and roughly 38 percent claiming "very often," beating out kids' toys such as dolls and action figures, arts and crafts, and construction-based toys, which all had a roughly 50 percent usage rate on the poll. Gaming consoles were at 50 percent, and other simple children's toys such as vehicles, puzzles, and board games were closer to 40 percent, http://mcgrc.com/wp-content/uploads/2014/02/MCGRC_Digital-Kids-Presentation_0220142.pdf.

65. C. M. Tennessen and B. Cimprich, "Views to nature: effects on attention," *Journal of Environmental Psychology* 15, no. 1 (1995), 77–85, http://www.sciencedirect.com/science/article/pii/0272494495900160.

66. R. A. Atchley, D. L. Strayer, and P. Atchley, "Creativity in the wild: improving creative reasoning through immersion in natural settings," *PLoS One* 7, no. 12 (2012), e51474, http://www.plosone.org/article/info%3Adoi%2F10.1371%2Fjournal.pone.0051474.

67. T. Hartig, M. Mang, and G. W. Evans, "Restorative effects of natural environment experiences," *Environment and Behavior* 23, no. 1 (1991), 3–26, http://eab.sagepub.com/content/23/1/3.abstract.

68. Terrapin Bright Green LLC, "The Economics of Biophilia: Why Designing with Nature in Mind Makes Financial Sense," white paper (New York: Terrapin Bright Green, 2012), 13, http://www.terrapinbrightgreen.com/downloads/The%20Economics%20of%20Biophilia_Terrapin%20Bright%20Green%202012e.pdf.

69. S. Kaplan, "The restorative benefits of nature: toward an integrative framework," *Journal of Environmental Psychology* 15, no. 3 (September 1995), 169–82, http://willsull.net/resources/KaplanS1995.pdf.

70. M. Berman, J. Jonides, and S. Kaplan, "The cognitive benefits of interacting with nature," *Psychological Science* 19, no. 12 (December 2008), 1207–12, http://pss.sagepub.com/content/19/12/1207.abstract.

71. A recent *Buildings* magazine article notes how even green building approaches can degrade acoustics. "Acoustics: The Biggest Complaint in LEED-Certified Office Buildings," http://www.buildings.com/article-details/articleid/14557/title/acoustics-the-biggest-complaint-in-leed-certified-office-buildings.aspx.

72. Study described by Catherine Franssen at Blue Mind 2, 5 June 2012.

73. C. Zhong, A. Dijksterhuis, A. Galinsky, "The Merits of Unconscious Thought in Creativity," *Psychological Science* 19, no. 9 (September 2008), 912–18, doi: 10.1111/j.1467-9280.2008.02176.x.

74. B. Baird, J. Smallwood, M. D. Mrazek, J. W. Y. Kam, M. S. Franklin, and J. W. Schooler, "Inspired by distraction: mind wandering facilitates creative incubation," *Psychological Science* 23, no. 10 (October 2012), 1117–22, http://pss.sagepub.com/content/23/10/1117.

75. Konrad Lorenz, *King Solomon's Ring: New Light on Animal Ways* (New York: Penguin, 1997; first published by Thomas Y. Crowell, 1952), 15.

76. Mihaly Csikszentmihalyi, "Finding flow," excerpted from *Finding Flow: The Psychology of Engagement with Everyday Life* (New York: Basic Books, 1997), in *Psychology Today*, 1 July 1997, http://www.psychologytoday.com/articles/199707/finding-flow.

77. Steven Kotler, *West of Jesus* (New York: Bloomsbury, 2006), 138–39.

78. Janata is more apt to use the term "groove" than "flow" when referring to that particular neurological response to music.

79. Janata has shown that in musical flow we are engaging both the directed attention-focus networks of the brain *and* the positive emotion/memory/self-reflective default-mode network. "Even though these two networks seem to operate in opposition, if you think of situations we enjoy, they necessarily involve both," he says.

80. Interview with Ellen Langer by Alison Beard, "Mindfulness in the Age of Complexity," *Harvard Business Review*, March 2014.

81. Matt Richtel, "Outdoors and out of reach, studying the brain," *The New York Times*, 15 August 2010, http://www.nytimes.com/2010/08/16/technology/16brain.html?pagewanted=all&_r=0.

82. Quoted in Richard Louv, *The Nature Principle: Human Restoration and the End of Nature-Deficit Disorder* (Chapel Hill, NC: Algonquin, 2011), 36.

83. R. W. Emerson, *The Selected Writings of Ralph Waldo Emerson*, ed. Brooks Atkinson (New York: Modern Library, 1964), 901.

Chapter 8: Only Connect

1. "MIDWAY: A Message from the Gyre, a short film by Chris Jordan," http://vimeo.com/25563376.

2. Matthieu Ricard is a Buddhist monk who left a scientific career as a molecular biologist in France to study Buddhism in the Himalayas more than forty years ago. He's been the French interpreter for His Holiness the Dalai Lama since 1989. Matthieu donates the proceeds from his work and much of his time to thirty humanitarian projects in Asia.

3. The phrase "implanted with electrodes" likely activated your empathy for macaques!

4. J. M. Kilner, A. Neal, N. Weiskopf, K. J. Friston, and C. D. Frith, "Evidence of mirror neurons in human inferior frontal gyrus," *The Journal of Neuroscience* 29, no. 32 (13 August 2009), 10153–59, http://www.jneurosci.org/content/29/32/10153.short.

5. Siegel, *Mindsight*, 61 (see chap. 3, n. 15).

6. U. Dimberg and M. Thunberg, "Empathy, emotional contagion, and rapid facial reactions to angry and happy facial expressions," *PsyCh Journal* 1, no. 2 (2012), 118–27, http://onlinelibrary.wiley.com/doi/10.1002/pchj.4/abstract.

7. C. Nicholson, "Q&A: Art Glenberg, on how the body affects the mind," *Smart Planet* 09 (8 March 2013), http://www.smartplanet.com/blog/pure-genius/qa-art-glenberg-on-how-the-body-affects-the-mind.

8. C. Lamm, J. Decety, and T. Singer, "Meta-analytic evidence for common and distinct neural networks associated with directly experienced pain and empathy for pain," *Neuroimage* 54, no. 3 (1 February 2011), 2492–502, http://www.ncbi.nlm.nih.gov/pubmed/20946964.

9. T. H. Huxley, "Goethe: Aphorisms on Nature," *Nature* 1, no. 1 (4 November 1869), 1, http://www.nature.com/nature/about/first/aphorisms.html.

10. A. J. Howell, R. L. Dopko, H. Passmore, and K. Buro, "Nature connectedness: associations with well-being and mindfulness," *Personality and Individual Differences* 15, no. 2 (July 2011), 166–71, http://www.sciencedirect.com/science/article/pii/S0191886911001711.

11. N. Weinstein, A. K. Przybylski, and R. M. Ryan, "Can nature make us more caring? Effects of immersion in nature on intrinsic aspirations and generosity," *Personality and Social Psychology Bulletin* 35 (October 2009), 1315–29, http://psp.sagepub.com/content/35/10/1315.abstract.

12. Gallagher, *The Power of Place*, 210 (see chap. 3, n. 14).

13. Sigurd F. Olson, *The Singing Wilderness* (New York: Alfred A. Knopf, 1956), 8.

14. Siegel, *Mindsight*, 232.

15. A. Juric, "Why we are called to the wild," *Inner Landscapes for Inner Explorers* blog, 14 June 2013, http://www.innerlandscapes.org/blog/2013/6/14/why-we-are-called-to-the-wild.html.

16. M. N. Shiota, D. Keltner, and A. Mossman, "The nature of awe: elicitors, appraisals, and effects on self-concept," *Cognition and Emotion* 21, no. 5 (2007), 944–63, http://greatergood.berkeley.edu/dacherkeltner/docs/shiota.2007.pdf.

17. S. R. Kellert with the assistance of V. Derr, *A National Study of Outdoor Wilderness Experience* (Washington, D.C.: National Fish and Wildlife Foundation, 1998), cited in P. Heintzman, "Spiritual outcomes of wilderness experience: a synthesis of recent social science research," *Park Science* 28, no. 90 (Winter 2011–12), 89–92.

18. Z. Josipovic, I. Dinstein, J. Weber, and D. J. Heeger, "Influence of meditation on anti-correlated networks in the brain," *Frontiers in Human Neuroscience* 5 (2012), 183, http://www.frontiersin.org/Journal/10.3389/fnhum.2011.00183/abstract.

19. L. M. Fredrickson and D. H. Anderson, "A qualitative exploration of the wilderness experience as a source of spiritual inspiration," *Journal of Environmental Psychology* 19 (1999), 21–39, http://www.sciencedirect.com/science/article/pii/S0272494498901104.

20. A. H. Maslow, Preface to *Religions, Values, and Peak-Experiences* (New York: Viking, 1970; reprint Penguin, 1994), xvi.

21. A. H. Maslow, *Toward a Psychology of Being* (New York: Start Publishing, Kindle edition, 2012), Kindle locations 1539–40.

22. See L. Smith, "A qualitative analysis of profound wildlife encounters," *Journal of Dissertation* 1, no. 1 (2007), http://www.scientificjournals.org/journals2007/articles/1194.pdf; E. Hoffman, "What was Maslow's view of peak-experiences?," *Psychology Today* blog, 4 September 2011, http://www.psychologytoday.com/blog/the-peak-experience/201109/what-was-maslows-view-peak-experiences; M. McDonald, S. Wearing, and J. Ponting, "The nature of peak experiences in Wilderness," *The Humanistic Psychologist* 37 (2009), 370–85, http://www.tandfonline.com/doi/abs/10.1080/08873260701828912?journalCode=hthp20#preview; and Fredrickson and Anderson, "A qualitative exploration."

23. Maslow, *Toward a Psychology of Being*, Kindle locations 1583–84.

24. J. Yogis, *Saltwater Buddha: A Surfer's Quest to Find Zen on the Sea* (Somerville, MA: Wisdom, 2009), 157.

25. K. S. Bricker and D. L. Kerstetter, "Symbolic uses of river recreation resources: whitewater boaters' special places on the South Fork of the American River," in *Water and People: Challenges at the Interface of Symbolic and Utilitarian Values,* S. F. McCool, R. J. Clark, and G. H. Stankey, eds. (Washington, D.C.: United States Department of Agriculture Forest Service, 2008), 161–62.

26. My cousin Matt Claybaugh has led an Ocean Wilderness Therapy program for the state of Hawaii serving youth at risk as well as adults. "In twenty years of taking people to sea, sail trainees from ages 4 to 80, I have yet to return when there wasn't a feeling of personal transformation shared by all," he told me.

27. Eastern religions use many ocean metaphors to describe the mind. The name Tenzin *Gyatso*, the fourteenth *Dalai* Lama, has the word for *ocean* in it twice (roughly meaning Ocean Wisdom, Ocean Teacher).

28. In 1929 Freud wrote about what his friend (Rolland, uncredited) believed was the "true source of religious sentiments": "It is a feeling which he would like to call a sensation of 'eternity,' a feeling as of something limitless, unbounded—as it were, 'oceanic.'" Sigmund Freud, *Civilization and Its Discontents*, trans. J. Strachey (New York: W. W. Norton, 1961), 11.

29. Euripides, "Iphigenia in Tauris," in *The Complete Greek Drama,* trans. R. Potter (New York: Random House, 1938), line 1193, http://www.perseus.tufts.edu/hopper/text?doc=Perseus%3Atext%3A1999.01.0112%3Acard%3D1153.

30. E. Woody, "People of the River—People of the Salmon, *Wana Thlama-Nusuxmí Tanánma,*" in *Water and People*, 183.

31. I. Foster, *Wilderness, a Spiritual Antidote to the Everyday: A Phenomenology of Spiritual Experiences in the Boundary Waters Canoe Area Wilderness,* master's thesis (Missoula: University of Montana, 2012), p. 65, http://etd.lib.umt.edu/theses/available/etd-06262012-124555/unrestricted/Foster.pdf.

32. Ibid., 147.

33. Ibid., 242.

34. Ibid., 150. Quote edited for style.

35. Ibid., 207. Quote edited for style.

36. M. Foster, "Bluefin Tuna Sells for Incredible Record $1.76 Million at Tokyo Fish Auction (VIDEO)," *Huffington Post Food,* 1 January 2014, http://www.huffingtonpost.com/2013/01/05/bluefin-tuna-sells-for-incredible-record-tokyo-fish-auction_n_2415722.html?view=screen.

37. http://www.catalinaop.com/ProductDetails.asp?ProductCode=b_w1, accessed 25 February 2014.

38. M. Foster.

39. S. K. Narula, "Sushinomics: how bluefin tuna became a million-dollar fish," *The Atlantic,* January 2014, http://www.theatlantic.com/international/archive/2014/01/sushinomics-how-bluefin-tuna-became-a-million-dollar-fish/282826.

40. J. Adelman and A. Mukai, "Tuna sold at record price is overfished, study says," *Bloomberg News,* 8 January 2013, http://www.bloomberg.com/news/2013-01-09/tuna-species-sold-at-record-price-faces-overfishing-study-says.html.

41. "Price of bluefin tuna falls at Tokyo auction," Associated Press, reported in *The Guardian,* 5 January 2014, http://www.theguardian.com/world/2014/jan/05/bluefin-tuna-tokyo-auction.

42. Ibid.

43. Along the 2.5-mile walk from the hotel to the venue to give a TED talk at a 2009 conference in Santa Monica about plastic pollution, I counted (and picked up) 346 pieces of plastic from the street and sidewalk.

44. http://www.sandiego.gov/thinkblue/news/videos.shtml, accessed 26 February 2014.

45. M. Roberts, "The touchy-feely (but totally scientific!) methods of Wallace J. Nichols," *Outside,* December 2011, http://www.outsideonline.com/outdoor-adventure/nature/The-Touchy-Feely-But-Totally-Scientific-Methods-Of-Wallace-J-Nichols.html.

46. B. Latané and J. M. Darley, "Group inhibition of bystander intervention in emergencies," *Journal of Personality and Social Psychology* 10, no. 3 (1968), 215–21, http://psych.princeton.edu/psychology/research/darley/pdfs/Group%20Inhibition.pdf.

47. M. van Vugt, "Are we hardwired to damage the environment?" *Psychology Today* blog, 20 June 2012, http://www.psychologytoday.com/blog/naturally-selected/201206/are-we-hardwired-damage-the-environment.

48. L. T. Harris and S. T. Fiske, "Social groups that elicit disgust are differentially processed in mPFC," *Social Cognitive and Affective Neuroscience* 2, no. 1 (2007), 45–51, http://intl-scan.oxfordjournals.org/content/2/1/45.full.

49. O. Klimecki, M. Ricard, and T. Singer, "Empathy versus compassion: lessons from 1st and 3rd person methods," in *Compassion: Bridging Practice and Science*, T. Singer and M. Bolz, eds. (Munich: Max Planck Society, 2013), 279.

50. For a summary of research on this topic, see F. Warneken, "The development of altruistic behavior: helping in children and chimpanzees," *Social Research* 80, no. 2 (Summer 2013), 431–42.

51. J. A. Grant, "Being with pain: a discussion of meditation-based analgesia," in *Compassion: Bridging Practice and Science*, 265.

52. Ashar et al., "Towards a neuroscience of compassion: a brain systems-based model and research agenda," in press (available at waterlab.colorado.edu/files/Ashar_et_al_Neurosci_of_Compassion_in_press.pdf), accessed 7 February 2014, p. 3.

53. C. A. Hutcherson, E. M. Seppala, and J. L. Gross, "Loving-kindness meditation increases social connectedness," *Emotion* 8, no. 5 (2008), 720–24.

54. Ibid., 1177.

55. D. DeSteno, "The morality of meditation: gray matter," *New York Times*, 5 July 2013, http://www.nytimes.com/2013/07/07/opinion/sunday/the-morality-of-meditation.html?_r=0.

56. Although in 1980 country music recording artist Eddie Rabbitt found fame and fortune by essentially repeating the phrase "I Love a Rainy Night": "Well, I love a rainy night, It's such a beautiful sight, I love to feel the rain on my face, taste the rain on my lips... Showers washed all my cares away, I wake up to a sunny day 'cos I love a rainy night, yeah, I love a rainy night!"

57. T. Roszak, "Where psyche meets Gaia," in *Ecopsychology: Restoring the Earth, Healing the Mind*, ed. T. Roszak (San Francisco: Sierra Club, 1995), 16.

58. C. Swain, "You've got to get wet," in *Oceans*, Jon Bowermaster, ed. (New York: PublicAffairs, 2010), 258.

59. Gallagher, *The Power of Place*, 214 (see chap. 3, n. 14).

60. J. Macy, "Working through environmental despair," in *Ecopsychology*, 253–54.

61. P. Lehner, "BP Oil disaster at one year: grasping the regional economic impacts," Switchboard: National Resources Defense Council Staff blog, 13 April 2011, http://switchboard.nrdc.org/blogs/plehner/gulf_fishing_and_tourism_indus.htm.

62. A. Casselman, "A year after the spill, 'unusual' rise in health problems," *National Geographic News,* 20 April 2011, http://news.national geographic.com/news/2011/04/110420-gulf-oil-spill-anniversary-health -mental-science-nation.

63. "New Clinical-Disaster Research Center," press release, University of Mississippi Department of Psychology, undated, http://psychology.olemiss .edu/psychology-team-conducts-research-on-bp-oil-spill-aftermath/, accessed 27 February 2014.

64. L. M. Grattan, S. Roberts, W. T. Mahan Jr., P. K. McLaughlin, W. S. Otwell, and J. G. Morris Jr., "The early psychological impacts of the Deepwater Horizon oil spill on Florida and Alabama communities," *Environmental Health Perspectives* 119 (2011), 838–43, http://ehp.niehs.nih.gov/1002915.

65. Abram, *The Spell of the Sensuous,* x (see chap. 4, n. 5).

66. D. Kahan, "Fixing the communications failure," *Nature* 463 (2010), 296–97.

67. For a full description of this lesson, see D. Goleman, L. Bennet, and Z. Bartow, *Ecoliterate: How Educators Are Cultivating Emotional, Social, and Ecological Intelligence* (New York: Jossey-Bass, 2012), 1–2.

68. Kirsten Weir, "Your cheating brain," *New Scientist,* 21 (March 2014), 35–37.

69. A. Grant, *Give and Take: A Revolutionary Approach to Success* (New York: Viking, 2013), 166.

70. K. J. Wyles, S. Pahl, M. White, S. Morris, D. Cracknell, and R. C. Thompson, "Towards a marine mindset: visiting an aquarium can improve attitudes and intentions regarding marine sustainability," *Visitor Studies* 16, no. 1 (2013), 95–110.

71. Bem is perhaps best known for his controversial parapsychology claims. I can't speak to that research, but his insight into other aspects of the mind and behavior were widely praised, and on this point I think he's incredibly perceptive.

72. T. D. Wilson, "We are what we do," in *This Explains Everything: Deep, Beautiful, and Elegant Theories of How the World Works,* John Brockman, ed. (New York: HarperCollins, 2013), 354–55.

73. Quoted in Gallagher, *The Power of Place,* 216 (see chap. 3, n. 14).

74. David Foster Wallace, *A Supposedly Fun Thing I'll Never Do Again* (New York: Little, Brown, 1997), 262.

75. I. Keskinen, "Fear of the water and how to overcome it," paper presented at the Australian Swimming Coaches and Teachers Association (ASCTA) Convention, Broadbeach, Australia, May 2000, http://users.jyu.fi/~ikeskine/artikkeli1.htm.

76. R. Louv, *The Nature Principle*, 44–45 (see chap. 7, n. 76).

77. David Gelles, "The Mind Business," *Financial Times*, 24 August 2012.

Chapter 9: A Million Blue Marbles

1. LeBaron Meyers is vice president for strategic partnerships at Urban-Daddy, a company that makes an extremely popular smartphone app that quickly guides undecided urbanites to the hippest nightlife and eatery options in cities around the world, based on self-ranking of the user's instantaneous, changing interests and mood. She called that glass blue marble a "killer app."

2. The sequels to James Cameron's box-office-record-setting film *Avatar* are water themed, but it's the "King of the World" scene in *Titanic*, with a young Leonardo DiCaprio in the bow of the ship, arms outstretched and face to the sky, that stands out as one of the ultimate cinematographic Blue Mind moments, http://www.imdb.com/video/imdb/vi2676989977.

3. Ben Freiman is a high school senior who was born in Monterey, California, and never truly left the sea behind. He was recognized as an Ocean Hero by Save Our Shores for his tireless work to keep our coast clean, and gave a speech at his bar mitzvah to spread ocean awareness, which led to his attendance at the first Blue Mind conference. Since then, he has taken a high school marine biology course and hopes to continue studying in this field and broadening the world's appreciation of the gift of the ocean.

4. A. Reinert, "The blue marble shot: our first complete photograph of Earth," *The Atlantic*, 12 April 2011, http://www.theatlantic.com/technology/archive/2011/04/the-blue-marble-shot-our-first-complete-photograph-of-earth/237167.

5. Carl Sagan, *Pale Blue Dot: A Vision of the Human Future in Space* (New York: Random House, 1994), 6, 7.

6. Giordano Bruno was an itinerant, rebellious, misfit, and insatiably curious sixteenth-century Italian Dominican friar, philosopher, mathematician, and poet known most for his cosmological theories. Among other ideas, he proposed that the sun was just a star moving in space and that the universe included an infinite number of worlds populated by intelligent beings. For these and other heretical ideas Bruno was tried, found guilty, and in 1600 burned at the stake by the Roman Inquisition. By our modern definition Bruno isn't considered to have been a scientist, although later scholars and commentators regarded him as a martyr for free thought and modern scientific ideas. Neil deGrasse Tyson included a depiction of Bruno's story in the remake of the PBS series *Cosmos*, from which this quote was sourced.

Index

About the Author

Wallace "J." Nichols, Ph.D., is a research associate at the California Academy of Sciences and cofounder-codirector of Ocean Revolution, SEE the WILD, and LiVBLUE. He lives in California with his wife, Dana, and two daughters.

BACK BAY · READERS' PICK

Reading Group Guide

BLUE MIND

by

WALLACE J. NICHOLS

An online version of this reading group guide is available at littlebrown.com.

Go Deep: A Readers' Guide to *Blue Mind*

by Jamie K. Reaser

> "Blue Mind *is, deep down, about human curiosity, knowing ourselves more and better.*"
>
> — Céline Cousteau

I've twice had the honor of being in the audience while Wallace J. Nichols ("J") spoke on the topic of Blue Mind. The first time was in 2013, at London's Royal Geographical Society, when he joined two other EarthWatch lecture series panelists in exploring "Why Emotion Matters in Conservation Science." The second opportunity occurred in the fall of 2014, when he addressed students and faculty in the University of Virginia School of Architecture. My impression of him was the same on both occasions: this is a man in love.

It would be hard to say exactly what J is in love with—the list might be quite long. Stories about his two daughters, wife, and father will cause tears to well up in his eyes. Watch him while he talks about sea turtles or the people with whom he has shared a career in sea turtle conservation and you'll notice that his voice softens, his cheeks flush, and his throat constricts. His heart is choosing the words. And then, there is water—everything about water. To J, water is muse, spiritual teacher, and refuge.

J is not your average scientist. He is a wayfarer, a captain of the Earth Ship, charting the way for us to become more fully human by reclaiming and celebrating our relationship with water, the element in which our lives took form.

Blue Mind: The Surprising Science That Shows How Being Near, In, On, or Under Water Can Make You Happier, Healthier, More Connected, and Better at What You Do is more impassioned missive to humanity than technical treatise on the linkage between hydrogen and hydroxyl ions. J draws on recent scientific findings to explain the impacts that proximity to water can have on human cognition and physiology, including a sense of ease and greater connection with other people and the natural world. His personal stories—punctuated with wonderment and awe—inspire the reader to reflect on the depth of his or her own relationship with water.

On every page, J beckons: "Come in, the water is fine." He hopes that we will release our pathological grip on the shoreline. He hopes we will remember that we are water creatures (the average adult body is 50–65 percent water) and engage in a relationship with water that helps heal our personal and collective ills—the ailments of Western culture that result in toxification of this water planet.

Some 72 percent of the Earth is covered in water. Water-quality studies indicate that at least one-third of the freshwater bodies in the United States are polluted, and every year at least one-quarter of our beaches are closed due to pollution. Worldwide, approximately 50 percent of the groundwater is not suitable for consumption. Every minute, two to three children dic of a water-borne disease. Falling in love with water is a matter of survival.

And fall in love we should. To have a Blue Mind is to have an awakened heart. *Blue Mind* has inspired not just new ways of thinking about water, but also new ways of feeling toward and acting on behalf of water. *Blue Mind* has also pulled scientists in the fast-growing field of neuroconservation into a new wave of research on topics such as "the brain on water." I expect that the findings will leave us in awe.

Some scientists might think that J has lost his marbles; being openly emotional about one's research topic is often considered "unprofessional." J, however, gives his marbles away freely. Blue marbles are J's hallmark. Meet him and you are likely to walk away with an iridescent deep-blue glass orb. Look into it. Begin to reimagine the possible for water, for humans, for humanity.

What's your relationship with water? What could it become? Curious? This reader's guide is designed to help you move from the shoreline into deeper waters. Below I offer chapter-by-chapter questions and activities to enable you to explore how you think and feel about water. Many of the questions are drawn from readers like yourself—people who want to care more and do more for water.

Ready? Jump in. Then create ripples by sharing your responses to the questions and activities through social media (use the #bluemind hashtag) and by encouraging your friends and colleagues to get a Blue Mind. As J dives even further below the surface, you can follow his work by going to his website, wallacejnichols.org.

May you come to know yourself more and better.

Jamie K. Reaser has worked around the world as a conservation ecologist, environmental negotiator, trainer in applied Neurolinguistic Programming (NLP), and wilderness rites-of-passage guide. She is the author of several books, most recently, *Winter: Reflections by Snowlight*. She lives in the Blue Ridge Mountains of Virginia. Portions of this essay were adapted from a feature on J and *Blue Mind* that appeared in *The Wayfarer,* volume 3, issue 4, Winter 2014.

Questions and activities for reading groups

Foreword and Preface

Marine explorer and conservationist Jacques-Yves Cousteau observed that "people protect what they love." What do you think he meant by this? In what ways does love motivate you? What are you protective of?

Céline Cousteau and J agree that it is time to "explain the magic." What magical experiences have you had in or near water? How do you explain what happened? Give at least two examples of how scientific explanations of human experience have generated changes in how we live and understand our world.

J published *Blue Mind* in 2014. How might the book be different if it were written twenty years earlier? Fifty years earlier? How might a version of *Blue Mind* written fifty years from now differ from the original?

Describe your own life story in terms of a series of encounters with water. How might your life have been different in the absence of water? Consider the lives of people who inhabit parts of the world where water is a scarce or highly polluted resource.

Go Deeper: Share the story of your relationship with water through images and/or sound. Post it online and tag it #bluemind. Be sure to include love.

Chapter 1: Why Do We Love Water So Much?

J has had a long love affair with water. What elements of the other-than-human world have you courted and/or been courted by? When did you first encounter this beloved? Who were you with and what happened? How has this experience made you the person you are today?

Consider the significant moments in human lives that traditionally involve water. What is water's role at these times? Is it a literal or a metaphoric one? If metaphoric, what does water represent and how does this representation change across cultures? Why?

Think of someone who has a strong fear or dislike of water. Explore the probable sources of these emotions. What does the aversion to water prevent this person from experiencing? How might we do a better job at inviting people to "jump in"?

Go Deeper: J shares lines from a poem by Lisa Starr in which she writes, "Go see how it's been preparing forever for today." Write a poem about how water has been preparing forever for today for you.

Chapter 2: Water and the Brain: Neuroscience and Blue Mind

About the brain and the ocean, David Poeppel is quoted as saying, "We're drawn to their mysteries . . . we strive to find a language to describe them." What mysteries of the universe are you drawn to? What language do you use to describe them? Intellectual? Emotional?

What do you know about the functioning of your own brain? How has neuroscience influenced your life? What insights can neuroscience offer about your love affair with the other-than-human world (your "water")?

"The brain on water" is new as a topic of neuroscientific investigation. Given the intimate relationship that humans have with water, consider why neuroscientists did not focus on water until recently. Why do you think topics such as music, food, and meditation have garnered their attention before water?

Monitoring your brain function is likely to become as simple and common as monitoring your heart rate. What kind of information would you want to have? How would you use this information? If everyone could have this kind of information, how might the world change? What are your hopes? What are your concerns? How might this relate to water?

Go Deeper: Pick a topic that you are passionate about. View talks related to neuroscience on TED.com. Make a list of the insights that these talks provide with regard to the topic you chose. Explore the relationship between your brain and your heart in determining what you are passionate about and how you live these passions into the world.

Chapter 3: The Water Premium

J starts this chapter by describing the people he encountered while traveling along the coastal trail from Oregon to Mexico. What emotion did these people have in common? How do you feel when standing at the edge of a body of water? What expressions have you noticed on the faces of other people when they are near water? How does this differ depending on the activity you and they are engaged in?

People are willing to pay a premium to live or recreate near water. How does this information explain our challenge in protecting coastlines and other waterfronts? How could we use this observation to facilitate the restoration and protection of oceans and waterways?

Richard Louv writes that "sustainable happiness is…found in our relationship with place." What is your "sweet spot" relative to water? Is there a specific body of water that you are drawn to? Is it natural or human-made? Do you prefer to be near, in, on, or under water? Why?

In what ways do people use art (photographs, paintings, film, music, etc.) in order to be close to water? Can something that represents water have the same emotional impact as actual water? Why or why not? How have you brought water into your life in this way? How much were you willing to pay to do so?

Go Deeper: Visit a real-estate website (e.g., zillow.com) and look at waterfront properties (aka "the front row"). Imagine living in or recreating at some of them. Consider what your experience would be like. Now look at the second row of properties. Consider what it would be like to occupy the second row rather than the front row. What economic value would you place on the differences in experience? Compare this to the actual price differential between the rows. Explore what other ways you invest or would be wise to invest more in experiences and memories rather than things and possessions.

Chapter 4: The Senses, the Body, and "Big Blue"

When you're in the vicinity of water, what senses are you aware of? How does this change when you are in water? What about when you are submerged under water? How does this sensual awareness influence your emotions and your relationship with water?

Consider romance. How does romance relate to water? What romantic experiences have you had that involved water? Why do you think there is such a strong connection between water and romance?

Frequently, people describe their lives through phrases related to water. For example, we might say, "I'm in over my head" or "I'm going with the flow" or "I'm just floating for now." What are some other examples? What do these metaphoric sayings reveal about our inner and outer experiences of water?

Advances in modern technology are enabling people to indirectly experience water in ways that can be highly impactful on the visual and auditory senses (e.g., high-definition sound and picture). What is possible through these experiences and what is lacking? Consider the interplay between real and virtual environments. In what ways could apps, film, and photography create lasting memories, awe, and wonder? How could these technologies benefit water conservation?

Go Deeper: The next time you are in water (even if that means the shower or bathtub!), explore it with all of your senses (sight, hearing, touch, taste, and smell). Notice what you have never noticed about water before. Pay attention to the memories of other experiences in water that arise. Explore them, especially the emotions that are present. Make a vow to be even more attentive to your senses when you interact with water in the future (even if that's washing the dishes).

Chapter 5: Blue Mind at Work and Play

Swimming champion and avid waterman Bruckner Chase turned his passion for water into a meaningful career. In the US and other westernized societies people are often discouraged from integrating their passions into their work. We commonly refer to "work and play" or "work and hobbies" as separate activities. Is this true for you? How would your life be different if your career had been kick-started by your passions?

In what ways does water play a role in your job, field of study, hobbies, and recreational activities? How does water motivate you? What if water were no longer available for these activities? How would your motivation change? How would your life change?

Do you know someone with a "water addiction"? Are you a water addict? If so, in what way? Describe a healthy water addiction and an unhealthy version. How does water addiction foster or inhibit water conservation?

Go Deeper: J and his friend Chuy Lucero have pledged to gather their families on the shores of Mexico when they are great-grandfathers and raise a toast to the sea turtles they have dedicated their lives to protecting. What vows are you willing to make with regard to water and the species whose lives depend on it for their survival? Make one or more of these vows and share them publicly with others through social media using the #bluemind tag.

Chapter 6: Red Mind, Gray Mind, Blue Mind: The Health Benefits of Water

What role has water played in your health and well-being? How has water put your health at risk? In what ways has it been healing?

J provides his perspectives on Red Mind, Gray Mind, and Blue Mind. What do these concepts mean to you? Consider people you know. Which mind-set do they most orient toward? How does this change with context? How does your mind-set change with time and place? Why?

Red Mind has become a prominent mind-set in modern society. How is this beneficial? How is it detrimental? What direct experience do you have of the benefits and limitations of Red Mind? How can we use Blue Mind to influence Red Mind in society?

People seek out water for relaxation. What are some ways that you can boost Blue Mind in your everyday life? In what ways could you use Blue Mind to create a more relaxed and creative atmosphere for your family, friends, and colleagues?

Go Deeper: Water played an important role in J's healing following a tractor accident. Undoubtedly there are some ways in which you express Red Mind and/or Gray Mind that undermine your health and well-being. Identify Blue Mind practices that can help you thrive physically and psychologically, even spiritually. Adopt them.

Chapter 7: Blue Unified: Connection and Water

Some people turn to water for inspiration, while others easily get bored sitting on a beach. What has been your experience? What do you think explains the difference in people's reactions to water? How do you think this has changed in society over time? Why?

People who feel a strong affinity for water often describe their connection with it as something quite personal, even intimate. For this reason, people often seek privacy and solitude when communing with water. How have your private experiences in or near water been different from those in the vicinity of other people (e.g., at public swimming pools or beaches)? What do we lose as individuals and a society when privacy and solitude are not available to us?

How could artists help people build a stronger connection to water? What kinds of projects can you envision? What kinds of arts events, movements, or campaigns? How could neuroscientists and psychologists help people build a stronger connection to water? What new research question would you like them to pursue?

Go Deeper: Remember a time when you felt intimately connected to water. Name the emotions that were present. Make a list of your insights and beliefs. Compare this to the times just before and just after this experience. Looking back, consider the differences and reflect on how the intimate experience with water changed you in some way, small or large. Do something to express your gratitude for the positive effect that water had (e.g., clean up a waterway, make a donation to an organization focused on water conservation, give a copy of *Blue Mind* to someone who could benefit from a deeper connection to water).

Chapter 8: Only Connect

Water is integral to the creation myths of ancient civilizations worldwide. What role did water play in the stories told by your ancestors? What role did water play in their ceremonies and rituals? In what ways has this relationship with water been passed down through time? What has been lost through time?

Important social interactions often take place in or near water. What role does water play in connecting people? What has been true of this experience in your life? Imagine these social events without water. What's different? Why?

People often report getting their best ideas in the shower. Have you had this experience, or have you found other water contexts to be particularly good at connecting body and mind? Are there patterns to your experience? Why do you think water is so good for enabling us to connect with aspects of ourselves?

Consider your next opportunity to be near or in water. Whom will you take with you and why? Is your answer different if the experience is a natural body of water or an artificial water source? Why? How does your answer change according to your intent for interacting with water?

Go Deeper: Each of us has a personal story or creation myth that involves water. Consider what yours is. Express your myth through a written story or some form of deeply expressive art (e.g., dance, theater, painting, rhythm-based music). Share your myth through social media using the #bluemind tag.

Chapter 9: A Million Blue Marbles

J opens the final chapter with a quote from Marcel Proust that reads, "The real voyage of discovery consists not so much in seeking new territory, but possibly in having new sets of eyes." In what way has *Blue Mind* given you new eyes?

When astronauts describe the profoundly transformative experience of viewing the water planet from space, they often use expressions of awe and wonderment. In what way has awe been a part of your own transformative experiences with water? How could a sense of awe be better used to inspire water conservation?

How can you share Blue Mind ideas with like-minded people, as well as people who have not yet acquired a Blue Mind? How can you use your Blue Mind to help us better protect watery places and water quality? What will be your first step? When will you start?

Go Deeper: Get a bag of blue marbles. Play with them. Meditate with them. Do silly and profound things with them. Share them with friends, colleagues, and strangers who will become fast friends. Most important, gaze into them. Notice what this crystal ball has to say about your future and water's future and your future relationship with water. Blue marbles are available from J at bluemarbles.org and wherever he happens to be speaking on Blue Mind.